$4.83   7/8/19   Bw (2009)

The Seventh Landing

Going Back to the Moon, This Time to Stay

Written and illustrated by Michael Carroll

 Springer

Michael Carroll
6280 W. Chesnut Ave.
Littleton CO 80128
USA
cosmicart@stock-space-images.com
http://stock-space-images.com

ISBN 978-0-387-93880-6   e-ISBN 978-0-387-93881-3
DOI 10.1007/978-0-387-93881-3
Springer Dordrecht Heidelberg London New York

Library of Congress Control Number: 2009927129

Printed on acid-free paper

Springer is part of Springer Science+Business Media (www.springer.com)

"Scientists study the world as it is.
Engineers create the world that has never been."

—*Theodore Von Karman*

Acknowledgements

My thanks to Gary "SiloMan" Baker for Atlas ICBM photos, Tomomi Niizeki of JAXA, Bill Worman for NASCAR wisdom, Karl Dodenhoff for Soviet Moon landers, Ben Guenther for his Orlon space suit shot, Ivy Zhang at *National Geographic's* China office for help in obtaining Qin Xian'an's beautiful photo of the launch of Shenzhou VI, Daniel Peters of the Hal Leonard Corporation for musical help, Anatoly Zak for all things Russian-space, Ted Stryk and Don Mitchell for beautiful revamped space panoramas, to Marianne Dyson, John Hall, Dan Durda, Bryce Cox, Doug Haynes, critique buddies Rebecca Rowe and Brian Enke, Mom and Dad for input and inspiration, and my writing soul-mate, Caroline. Special thanks goes to Maury Solomon at Springer, who shared the vision of this book with me and saw it to fruition, and to the engineers, scientists and astronauts and Public Information Officers (with special commendation to Lynette Madison and Joan Underwood) who put up with my nagging!

About the Author

Artist/writer Michael Carroll has over 20 years experience as a science journalist, which has left him well-connected in the planetary science community. He has written articles and books on topics ranging from space to archaeology. His articles have appeared in *Popular Science, Astronomy, Sky and Telescope, Astronomy Now* (UK), and a host of children's magazines. His latest book, *Alien Volcanoes*, was published by Johns Hopkins University Press in the fall of 2007. His children's book *Dinosaurs* was a finalist for the ECPA's Gold Medallion award for excellence in publishing.

Carroll has done commissioned artwork for NASA and the Jet Propulsion Laboratory. His art has appeared in several hundred magazines throughout the world, including *National Geographic, Time, Smithsonian, Astronomy,* and others. One of his paintings is on the deck of the *Phoenix* Mars lander at the Martian north pole. Carroll is the 2006 recipient of the Lucien Rudaux Award for lifetime achievement in the Astronomical Arts, and is a Fellow of the International Association for the Astronomical Arts.

Foreword

In the early morning hours of July 16, 1969, I stood with several hundred NASA and contractor employees and journalists outside of the crew quarters at Kennedy Space Center. I watched as three young Americans, Neil Armstrong, Buzz Aldrin, and Michael Collins, emerged, clad in their white spacesuits. They walked by me, just a few feet away, to a waiting van that took them to Launch Pad 39A for the first step in their journey to the Moon. It was almost as if I had been present on August 3, 1492, as Christopher Columbus set sail from Spain to the New World. Almost four days later, Armstrong and Aldrin became the first humans to land on Earth's off-shore island.

I plan once again to be nearby, a decade or so from now, when four new explorers set off from Florida for humanity's seventh landing on the Moon. Just as was the case in 1969, that expedition will be a milestone in human history.

In *The Seventh Landing*, Michael Carroll gives us in well-chosen words and vivid images both concise background on the initial round of lunar exploration four decades ago and a clear sense of NASA's current plans or returning to the Moon. Although there are certain to be changes in the specifics of these plans as they mature, NASA is well along in designing the systems for getting people back to the lunar surface and for setting up a long duration human outpost there. Carroll has talked with many of those who are actively involved in this new effort, and he succeeds in communicating their almost palpable excitement as they lay the foundation for sustained human exploration beyond Earth orbit. He also interacted with representatives of the first generation of lunar explorers and finds them equally excited that the United States is finally heading back to the Moon, so many years after its first tentative exploratory journeys.

The U.S. return to the Moon will be a very different undertaking than was Project *Apollo*. Then, the goal—getting to the Moon before the Russians—was linked to the broad political contours of the U.S.-Soviet Cold War competition for world leadership. *Apollo* was a unilateral demonstration, peaceful in character, of American power. Maintaining American leadership is also a goal this time around, but in a very different political context. Many countries are now involved in space exploration, and the United States hopes that they will join a U.S.-led, but global in scope, undertaking.

The *Apollo* goal of being first to the Moon was achieved when Neil Armstrong took that first "small step for a man." There is no similar end point this time around. The seventh landing on the Moon will be just the first achievement in an open-ended effort to explore the Moon, and eventually Mars and beyond—an enterprise that will last for decades, even centuries.

Carroll not only describes the hardware that will carry astronauts to the Moon and allow them to stay there for extended periods. He also summarizes the many reasons why lunar exploration beyond that carried out during *Apollo*, the eventual exploitation of lunar resources, and using the Moon to prepare for journeys to Mars or other distant destinations are worthwhile objectives that should command the support of the American public and political leadership. The United States has not yet made a final decision to move forward with its human exploration plans; that decision requires the allocation of the funds freed up by retiring the space shuttle in the next few years to building the lunar lander and the launch vehicle for travel to the Moon. I certainly hope that the president who takes office in January 2009 agrees that a return to the Moon should be part of America's future.

NASA and thirteen other space agencies in 2006 began an effort to coordinate their space exploration plans, and in May 2007 the group issued a "Global Exploration Strategy." It is worth quoting: *"Opportunities like this come rarely. The human migration into space is still in its infancy. For the most part, we have remained just a few kilometers above Earth's surface—not much more than camping out in the backyard."* It is time to leave humanity's "backyard." Michael Carroll in *The Seventh Landing* tells us in words and pictures how that can happen. It is up to all of us to make real this vision of the future.

–John M. Logsdon

John M. Logsdon is Professor Emeritus of Political Science and International Affairs at George Washington University, Washington, D.C., and author of *The Decision to Go to the Moon: Project Apollo and the National Interest.* Currently he is the Charles A. Lindbergh Chair in Aerospace History at the National Air and Space Museum.

Contents

Introduction   Doing It Right xv

One   The First Explorers: Learning from History 2

Two   Getting There the Second Time Around 32

Three   Shackleton, the Home Site 62

Four   Robot-Human Combo Systems 78

Five   Scientific Reasons to Return 98

Six   Going to Mars 120

Afterword   To Boldly Stay 149

Chapter Notes 151

Appendix 1   The Evolving Space Program 155

Appendix 2   Moon Missions 163

Appendix 3   Mars and Asteroid/Comet Explorers 167

Index 171

Introduction

Doing It Right

July 20, 1969, marked the day that humanity first set foot on another world. In a mad, Cold-War dash to "get there first," America spent nearly a decade in the effort to put humans on the surface of the Moon and, as President John F. Kennedy had challenged, "return them safely to the Earth." On that summer day in 1969, Neil Armstrong became the first of a dozen people to explore the dusty plains and rolling mountains of Earth's nearest cosmic neighbor.

Visionaries drew plans to add permanence to *Apollo's* sorties. Some space architects dreamed of modules set down on the lunar plains to house visiting scientists. Others applied early 1970s technology to more ambitious missions to Mars. Although their Moon program had not succeeded, Soviet planners continued to put forth creative ideas for populating and exploring the lunar environment. But in the end, the world's advanced plans were only dreams. Although the National Aeronautics and Space Administration's (NASA's) lunar expeditions brought home a wealth of scientific booty that transformed planetary and terrestrial science, critics felt that *Apollo's* missions constituted a splash with no ripples, leaving behind not a foundation for future exploration but a heritage of flags and footprints.

The *Apollo* program, and NASA's human exploration program in general, fell victim to the perfect storm of bad circumstances. The financial black hole of Vietnam took its toll on many aspects of the American economy. Subtle currents in the world's political arena made competition with the Soviets less appealing. Social pressures at home and an awakening of the civil rights and environmental movements pulled national focus away from exploration. Ironically, lunar and planetary programs contributed a great deal to our understanding of Earth's environment, but the public of the 1970s and 1980s seemed better at compartmentalization than at holistic viewing. Although the quintessential image of the environmental movement—an *Apollo 8* snapshot of the Earth rising above the Moon—gave the green movement a focus and an icon, U. S. taxpayers saw the situation as either/or: spend money on healing the planet, or spend it on space exploration; invest in jobs, or in the Moon.[1]

1. *Apollo* fed hundreds of thousands of families, infusing American society with technology and a vibrant economy in the 1960s.

So the Moon was abandoned in 1972, but it will not remain so for long. The Moon is too tempting a resource, say proponents of exploration. Space historian John Logsdon calls it Earth's "off-shore island." Just a three-day journey away, its quarter-of-a-million-mile distance from Earth makes communication easy. With its proximity to Earth, the Moon provides a laboratory with which to understand our own world better. Exploration and technological challenges drive world economies, further medical and transportation arenas, and fire inventions that spill over into public use. The computer on which these words were written owes much of its heritage to *Apollo*-spawned technology, as does the cell phone that summons us and the microwave that heats our coffee. Space exploration has also served, historically, as a valuable tool in foreign relations. The partnership in scientific communities of planetary exploration during the latter stages of the Cold War served as a critical communications conduit between governments and continues to cement relationships between spacefaring nations.

Behind the diplomacy, practicalities, and spinoffs, something deeper beckons. Gunpowder-gray mountains glisten against velvet-black skies. Vast canyons drop precipitously from gently rolling plains. Titanic amphitheaters of rock bear silent witness to an ancient epoch of falling asteroids and comets, while lava plains hint at magma seas in eons past. Secrets are waiting there. Adventure is waiting there. So are new horizons—and unseen possibilities—for humankind.

The Moon stands today not so much devoid of life as abandoned by it. It's been thirty-seven years since people last trod the dusty plains of the Moon. Over the course of six landings, from 1969 to 1972, twelve men explored, four-wheeled, dug, and hiked across the lunar surface. Now, the United States envisions a seventh landing on the Moon. This time, it plans to stay.

Far more than flags and footprints, NASA has crafted a detailed blueprint to carry out this vision. The plan is called the Constellation Architecture. To accomplish its goals, designers are hard at work on a new family of launch vehicles, the most powerful in history. These advanced launchers borrow the best technology from the space shuttle, Saturn V, and other launch vehicle programs.

Harrison Schmitt samples a giant boulder in the mountains of Taurus Littrow during the last Apollo flight in 1972. (Photo courtesy NASA/JSC)

The *Orion* spacecraft forms the backbone of the program. *Orion* will carry four to six-people and is capable of trans-lunar travel. Other craft will ferry crews to and from the lunar surface from Moon orbit, and still others will supply the infrastructure for permanent settlements.

Compared with previous plans to return to the Moon, such as the Space Exploration Initiative of 1989, the Constellation program entails a long-term, methodical approach to exploration and settlement of the Moon and Mars. This time around, NASA has offered a lean, efficient, and well-reasoned version of former plans. The *Orion* spacecraft, replacement for the venerable shuttle program, is estimated to cost less to build and launch than either the shuttles or the proposed craft in earlier scenarios. The new blueprint to the stars spreads costs while building, step-by-step, on technology and lessons learned from setting up a base on the Moon.

Logo of NASA's Constellation Moon Project (Photo courtesy NASA/JSC)

NASA will have a great deal of company. Plans for future lunar exploration are being drawn up by Europe, Japan, Russia, China, and India. The real question facing NASA is this: *What part will we play*? Will the United States be observers on the sidelines, or energetic international partners on this new journey for humanity?

Despite today's challenges, humanity once again looks to the heavens, casting its eyes toward the Moon and farther. NASA's Constellation plans go far beyond Earth's nearest planetary neighbor. The launch vehicles of the Constellation Architecture have the capability to transport large loads into interplanetary space, carrying their cargoes across the 50-million-mile void to Mars. Eventually, the lessons learned on the Moon's outpost at Shackleton Crater promise to teach us how to live—permanently—on one of the most Earthlike worlds in our Solar System. With its vast natural resources and keys to planetary evolution and history, Mars beckons.

Although specific hardware and mission details will be in flux for some time, the overarching goals, strategies, and inspiration for the seventh landing will not change. In the chapters to come, we'll explore the basics of NASA's current strategy to return humans to the Moon: the ideas, inspiration, and technological contributions of the world's spacefaring community; details about setting up camp in the hostile lunar environment; reasons for a return, from science to society; and, finally, alternatives to Constellation and the ongoing work to get humans to Mars and beyond. But to understand the Constellation's unique approach for a return to the Moon, one must first understand what has come before.

"The moon, like a flower
in heaven's high bower,
with silent delight
sits and smiles on the night."
William Blake, English poet

"Treading the soil of the moon, palpating its pebbles, tasting the panic
and splendor of the event, feeling in the pit of one's stomach the
separation from terra . . . these form the most romantic sensation an
explorer has ever known . . . this is the only thing I can say about the
matter. The utilitarian results do not interest me."
Vladimir Nabokov, Russian-born American novelist

"Many years ago the great British explorer George Mallory, who was
to die on Mount Everest, was asked why did he want to climb it.
He said, 'Because it is there.'

"Well, space is there, and we're going to climb it, and the moon and
the planets are there, and new hopes for knowledge and peace are
there. And, therefore, as we set sail we ask God's blessing on the most
hazardous and dangerous and greatest adventure on which man has
ever embarked."
U. S. President John F. Kennedy (from speech given on September 12, 1962)

Chapter One

The First Explorers: Learning from History

The Soviet Luna 9 (art by author)

It was to be the first vehicle of its kind: a robot lander crafted to settle on the surface of Earth's Moon. Luna 8, pride of the Soviet Union, would be a window into another world, an illuminator of mysteries. For three days in 1965, the Moon swelled from a gibbous orb to a vast gray wall as the craft sailed toward Earth's sibling. Crater rims stood out against the dark sands, capped by rugged boulders. The beachball-sized craft raced through the translunar emptiness, ready. Folded around the lander, airbags were poised to fill with gas to cushion Luna's initial contact with the surface. Its petals

M. Carroll, *The Seventh Landing*, DOI 10.1007/978-0-387-93881-3_1,
© Springer Science+Business Media, LLC 2009

pulled against springs, waiting to open like a flower on the lunar surface, with cameras ready to unveil the unknown world to waiting scientists back in Moscow.

But it was not to be. Luna 8's retrorockets failed, and the craft slammed into the Moon's Sea of Clouds at 36,000 mph. Its failure was by no means the first, nor would it be the last.

Spacefaring nations of the world are drawing plans for lunar exploration, both robotic and human. A groundswell of support for a return to the Moon is building in the halls of universities and the laboratories of aerospace companies. The world's designers and dreamers are putting ideas to paper, and even to hardware. The details of how humankind will return to the Moon are as yet unknown, but a picture is coming into focus. Studies in Europe and Asia differ from those in the United States, but they share much in common, too. The most advanced studies, now taking physical form in aerospace factories across the United States, come from NASA's project Constellation. The Constellation Architecture, the blueprint for a permanent return to the Moon, serves as a good case study for how people will set up permanent residence on Earth's nearest neighbor.

Constellation is an overarching plan to return humans to the Moon as a permanent presence. To do so, NASA must forge a transportation system that affords permanent, low-cost access to space, a system capable of lifting large payloads into Earth orbit and to the Moon. For lunar missions, Constellation envisions two launches. The first, carried out by the Ares 1 booster, lofts an *Orion* Crew Exploration Vehicle (CEV) into orbit. The second launch sends a giant stage into Earth orbit, topped by the moonship Altair. After *Orion* links up with Altair, the upper stage will send the entire stack to the Moon. The plan has deep roots in engineering, flight experience, and history.

NASA's Constellation Architecture did not develop in a vacuum. Each new human endeavor is linked to the things that came before. Ancient Chinese explorers dreamed of trips to the moon on couches backed by solid rockets, technology that reappears each year on Guy Fawkes Day or the Fourth of July. A full thirty years before the first humans took those giant leaps for mankind, engineers and prescient draftsmen at the British Interplanetary Society drew plans involving solid rocket transports to the Moon. An updated 1947 study designed a liquid-fuelled moonship with four legs, hauntingly similar to the real future plans of American and Soviet aerospace designers.

As engineers and designers approach the Moon this time around, they ask themselves: What scenarios worked before? Should we revisit designs that came from Soviet and American lunar programs of the past? What can we learn from the booster technologies of the Cold War era, and what can we borrow from programs such as the Delta, Atlas, space shuttle, and International Space Station? In short, what worked...and what did not? History has rich lessons to teach us in the arenas of technology and exploration. It also has some fascinating sagas—and more than a few cautionary tales—to tell of our first close glimpses of Earth's nearest neighbor in space.

THE FIRST EXPLORERS

The Moon was the obvious target for early space explorers. The first Moon-bound spacecraft were tooled by Soviet and U. S. engineers. Visionaries and scientists had a multitude of clever and diverse ideas about how best to send a robotic emissary to our nearest cosmic neighbor. It seemed, at first blush, to be a simple prospect: strap a handful of science instruments to the top of a booster, hurl them toward the Moon, and wait for the pictures and measurements to stream back.

The Jupiter-bound Galileo *spacecraft snapped this beautiful image of Earth's nearest neighbor in space. This view is from above the lunar north pole. (Photo courtesy of NASA/JPL)*

But the 240 million miles (380 million km) spanning the Earth-Moon distance proved a daunting route to traverse. Aside from the vacuum of space, which could suck the life from conventional terrestrial equipment, temperatures in the sunlight reached 100° C (212° F), while the shadowed side of a spacecraft plunged to −173° C (−279° F). And what of communication? Engineers had no knowledge of how radio signals would perform across the void.

Early lunar explorers were powered by solid rockets, kerosene, and hydrazine. But they were also fueled by political forces. A sense of urgency permeated the decades of the fifties and sixties. The generation that saw its parents fight the Second World War now faced a new kind of threat. This war was a silent one, lacking outward explosions and obvious gunfire, but holding a tremendous sense of tension. Post-World War II peace and prosperity gave way, gradually and inexorably, to a balanced coexistence between two great powers. The uneasy equilibrium had been carved out between the eastern nations of the communist block and those nations in the west, represented for the most part by democratic governments. Like wars before it, this was a war of ideologies and ideas. To be sure, there was painful, violent physical conflict in places such as Korea and Vietnam, but a great percentage of Cold War resources were directed toward building and testing the biggest, most advanced weapons of mass destruction, high-tech sabers to be rattled over the heads of a nervous world. Nuclear warheads topped intercontinental ballistic missiles, and somewhere, at all times, someone had

Warhead-tipped intercontinental ballistic missiles became the workhorse of the early human space program. Atlas boosters similar to this one carried the first Mercury *astronauts into orbit. (Image courtesy of SiloWorld.)*

their finger on the trigger, poised to call radioactive fire down from the sky. Ironically, the threat of mutual destruction lent a sort of quiet constancy to the planet.

Within the nightmarish symmetry, a new generation looked to the sky. It was a generation that bridged two worlds. The first was a world of dreams, a place inhabited by Tiger Women of the Moon, four-armed green Martians, and space explorers wearing gravity belts and aviator caps. The second world was a place every bit as wondrous, a realm of real worlds, planets, and moons seething with volcanoes, awash in methane oceans, blanketed by cloud depths of crushing pressures laced with multi-colored poisons, a world where it just might be possible for people to explore strange new places. Political agendas aside, space offered a place for dreams to soar above the long, dark nights and tension-filled days of the Cold War.

In fact, it was Cold War politics that truly launched lunar exploration. When U. S. President John F. Kennedy challenged the United States to get humans to the Moon before the end of 1969, only the Soviets had actually sent a person into orbit. The Soviets had larger, more powerful rockets for their warheads, and rockets could be used to ferry people into the cosmos. If space was the new high frontier, the Soviets were making the first land claims. They seemed to have a distinct edge, and Kennedy wanted to level the launching field. "We choose to go to the Moon," he told an audience at Rice University in Texas. "We choose to go to the Moon in this decade and do the other things, not because they are easy, but because they are hard."

The remarkable statements Kennedy made on that September 12th in 1962 proved to be absolutely right. Going to the Moon would be hard. Vast resources would be invested. False starts and redesigns would test even the most seasoned engineers, scientists, and astronauts. Many people on both sides of the world would triumph during this great race to the Moon, but others would lose their lives in the cause. And within the designs and plans and explorations, the seeds of NASA's Constellation Architecture were sown.

From 1959 to 1965, the Soviets and Americans tossed at least 18 robotic spacecraft toward the Moon. Early successes included the Soviet *Luna 2,* which impacted the Moon, Luna 3[2], which snapped the first images of the far side of the Moon, and the American craft *Ranger 7,* which transmitted 4,300 images of the lunar surface before its planned impact in Mare Cognitum. But to really understand a place, a spacecraft has to get down to the surface, look around on a human scale, and get a little Moon dirt on it.

2. October 7, 1959

Airbags and Retrorockets

Engineers designed the first lunar probes to either fly by the Moon or crash into it. But landing on Earth's closest celestial neighbor is far more complex. A spacecraft approaches the Moon at a speed of roughly 1.6 miles (2.6 km) per second. When a craft returns to Earth from space, atmospheric friction slows the incoming vehicle considerably. Speed is transformed into heat as the craft reenters. Parachutes can deliver a payload to the ground gently.

The successful Mars Exploration Rovers landed with the use of airbags such as the ones seen in a test at left (note the figure on the ladder for scale). At right, the rover Spirit photographed its landing platform, with airbags retracted beneath. (Photos courtesy of NASA/JPL.)

The Moon offers no atmospheric drag for an initial slowing, and no air for parachutes. Any landing requires a large stock of propellant to slow the craft. Each *Luna* lander was ferried across the void aboard a carrier stage weighing 1.8 tons, much of which was fuel for reducing speed. After a three-day flight from Earth, at a range of 45 miles (75 km) from its destination, the carrier stage jettisoned radar and navigation equipment that was no longer needed for final stages of flight. The craft's main engine then began to slow the *Luna* spacecraft. The flight path carried the lander to the Moon's equatorial region, as the direct route would then result in the craft approaching the lunar surface vertically. Engines brought the craft to a full stop a few meters above the surface. A small pole sensed the surface as the stage fell to the ground. When the pole made contact, the *Luna*—perched atop the stage—popped from the stage and bounced to rest. But tests showed that even in the Moon's weak gravity, 1/6 that of Earth's, the delicate instruments would be damaged. So designers devised a dual-airbag system to cushion the fall. Once on the ground, the bottom-heavy spherical craft rolled to a stop. The outside covering split into four petals that righted the *Luna*. Over three decades later, similar airbag landing systems were successfully utilized on Mars with the U.S. *Pathfinder* craft and Mars Exploration Rover vehicles.

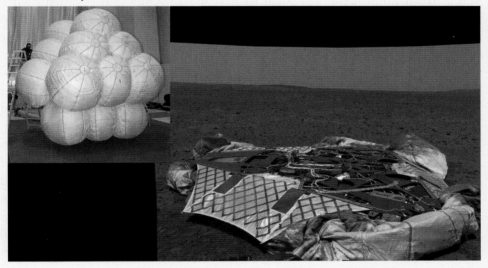

Finally, in the wake of a string of debris-filled craters left by its precursors, the Soviet *Luna 9* touched down softly on the lunar plains on February 3, 1966. The 99 kg (218 pound) craft was the size of a beach ball, 58 cm in diameter. Just 46 seconds before it hit the surface, the craft fired retro rockets, coming to a complete stop above the lunar plain. Airbags inflated only seconds before the craft hit the surface. The metallic ball rolled to a stop, opened four flower-like petals to right itself, and deployed spring-loaded antennae. A video camera set to work immediately, turning in place and using a scanning mirror to image its surroundings in the Ocean of Storms.

The first major discovery concerning the mission was that the craft did not sink into oblivion. Some researchers had predicted that the Moon was covered in a layer of dust deep enough to swallow any landing craft. This would have

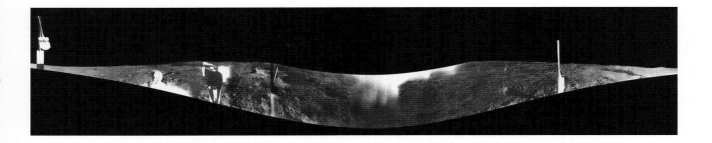

The first panorama from the surface of the Moon, returned by the Soviet Luna 9. *(Photo courtesy of Ted Stryk.)*

presented serious problems for human explorers. The prospect conjured up visions of Moonwalkers with snowshoes and hip-waders. But the surface of Oceanus Procellarum proved capable of firmly holding up the *Luna*.

The secretive Soviet Union announced its success immediately, but Moscow did not release photos for days. Instead, the first images came through a circuitous journey ending in England. Britain's Jodrell Bank Observatory intercepted the signals. Staff astronomer Sir Bernard Lovell noticed that the signal from *Luna 9* was identical to that used by newspapers to transmit images. Editors at the British tabloid newspaper *The Daily Express* rushed an image converter to the observatory and transformed Luna's signals into the first published panorama of the Moon's surface. Thanks to the Soviet success, the world witnessed a foreboding vista of rocks, craters, and deep shadows. *Luna 9's* camera even glimpsed some of its own hardware. The ghostly forms glistening in the brutal sunlight gave Western observers first-hand clues about Soviet technology.

Some mystery surrounds the series of events immediately following landfall. Why did the Soviets delay the release of *Luna's* images? The spectacular panoramas must certainly have oiled the Soviet propaganda machine and bolstered Russian political currency. Some analysts suggest that Semyon Lavochkin (1900–1960), designer of the *Luna* spacecraft, purposely outfitted it with universal imaging equipment, intending receivers at Jodrell Bank—who had the best equipment in the world—to decode and publish the information. In this way, the lower quality of Soviet technology would not come to light. Others proposed that the move was a political one, illuminating the struggle between internal forces within Soviet government. Some in the Soviet Union believed that the military arm of the government had used Yuri Gagarin, the first human to orbit Earth, as a shameless propaganda tool, and wanted to avoid a rerun by releasing data readily decodable by the west. In this way, the grinding Soviet propaganda machinery would not have time to twist the triumph into a military one.

Whatever the motives, the next three days saw a total of three complete panoramas of the lunar surface in seven communications sessions. On February 6, 1966, with its batteries depleted, *Luna 9*—first emissary from Earth—fell silent.

The first soft landing by a U. S. probe took place four months later. NASA's *Surveyor 1* rocketed to a safe landing on the plains of Oceanus Procellarum on June 2, 1966. Rather than using airbags, *Surveyor* carried out a powered

Lunar surface image returned by Surveyor.

descent using rocket engines, an approach that would not only serve *Apollo's* landings well but would provide the model for NASA's future Constellation vehicles. Thanks to its solar power, *Surveyor* lasted until July 14, returning 11,240 surface images over two lunar days. The craft's electronics finally succumbed to the harsh temperatures of lunar nights. Four successful siblings[3] followed *Surveyor 1*, each gaining in complexity. Some carried temperature and radiation sensors, while others were equipped with soil chemical analyzers and arms that scooped and dug in the lunar sands. All experiments were geared toward determining the safety of sending astronauts to the surface, and engineers targeted all Surveyors but one to land in areas under consideration for future *Apollo* landing sites.[4]

The triumphant *Luna 9* also had company. After several *Luna* orbiters, *Luna 13* landed safely in December of the same year and transmitted five full panoramas, along with data on soil properties.

MORE ADVANCED ORBITERS

A series of lunar orbiters from both the United States and the Soviet Union mapped the Moon's cratered surface from pole to pole, searching for the ultimate goal—a place safe enough to send humans. NASA's *Lunar Orbiters 1, 2,* and *3* were placed into low-inclination orbits so that they could image proposed *Apollo* landing sites in detail; all *Apollo* sites were fairly near the equator. *Lunar Orbiter 1* began circling the Moon in August of 1966. It was followed at three-month intervals by *Lunar Orbiters 2* and *3*. *Lunar Orbiters 4* and *5* settled into higher altitude polar orbits, enabling mapmakers to chart the entire lunar globe. These missions also had broader scientific goals. *Lunar Orbiter 5* ended its mission in August of 1967. For NASA, the stage was set to send the first human explorers to another world.

3. *Surveyor 3, 5, 6,* and *7*.

4. *Surveyor 7* landed in the southern highlands near the rim of Tycho crater. The region was thought too dangerous for a piloted landing and was too far from the equator for *Apollo's* limited fuel.

The Grand Plan from an American Perspective

The Soviet and American approaches to piloted Moon travel differed considerably. NASA's plans, advocated by chief engineer Werner Von Braun, involved the use of two vehicles to get to the Moon. Designers considered four scenarios:

1. Direct: A spacecraft would be sent on a *direct ascent* to the Moon, land, take off and return all in the same massive vehicle. This approach would require a monstrously powerful booster.

2. Earth Orbit Rendezvous (EOR): A giant spacecraft similar to the direct ascent craft would meet in low Earth orbit with a propulsion unit that would take the craft to the Moon. This scenario would make use of two Saturn V boosters.

3. Lunar Surface Rendezvous: Another plan called for two spacecraft to be launched Moonward. The first would be unmanned and fully automated, and would ferry fuel to the lunar landing site. The second craft would carry a crew to land on the Moon near the first craft, and transfer fuel for the return trip.

4. Lunar Orbit Rendezvous (LRO): The fourth plan involved one launch that would send two spacecraft into lunar orbit. In orbit, the two spacecraft would separate, with one landing and returning to rendezvous with the waiting "mother" ship that would bring the crew back to Earth. This plan required astronauts to link up while on the far side of the Moon, out of communications with Earth.

It seemed risky at a time when no spacecraft had linked with another. Though EOR was initially favored by the American teams, Lunar Orbit Rendezvous, championed by a Langley Center engineer named John Houbolt, ultimately was selected for *Apollo*. One launch of the Saturn V lofted both the *Apollo* Command and Service Module (CSM) and the Lunar Module (LM, pronounced "Lem"). The LM was plucked from the third stage and joined nose-to-nose with the Command and Service Module en route to the Moon. The main engine of the CSM slowed the entire group into lunar orbit. Two astronauts transferred into the LM for descent to the lunar surface while the third crewmember, in *Apollo,* circled in lunar orbit overhead. After serving as living quarters and base for the lunar explorers, the LM launched from the surface on its ascent stage, leaving the descent stage behind as a launch platform. In orbit, the two Moonwalkers returned to join their fellow astronaut in the Command Module, jettisoned the LM, and returned to Earth using the main engine on the CSM. Only the Command Module would make it all the way back, as the Service Module was jettisoned before reentry. The advantage of this plan over others was that it required far less fuel.

NASA's new Constellation Architecture shares commonality with both the EOR and LRO scenarios, but uses two launches to provide enough supplies for four crewmembers (in contrast to *Apollo's* two) to explore the lunar surface.

The Soviet Moon program was still in full swing. Soviet cosmonauts had racked up many space "firsts," including the first human in orbit and the first space walk, or EVA (Extra Vehicular Activity). The Soviets appeared to be preparing to unveil a new spacecraft as well. Western analysts wondered what they were up to.

The Soviets' unmanned program made the West as nervous as their piloted flights did. In parallel to the *Luna* program, the Soviets carried out the mysterious *Zond* probe missions. The *Zond* program baffled Western scientists for some time, as it seemed to go in many directions. *Zond 1* was launched in the direction of Venus, while *Zond 2* headed off toward Mars. Neither of these missions appears to have been successful. *Zond 3* flew past the Moon in July of 1965, imaging areas of the far side not seen by *Luna 3*. It returned 26 images of the far side, three in the ultraviolet spectrum. *Zond 3* may have been a twin craft designed to fly to Mars at the same time as *Zond 2.* For some reason, the craft missed the launch window, becoming

The First Space-Walker

When it came time for the Soviets to discuss who would be the first cosmonaut to walk on the Moon, the name that kept surfacing was Alexei Leonov. General Leonov was the first person to leave the pressurized safety of a spacecraft to venture out into the vacuum of space. His daring eleven-minute space walk proved that people could live and work in the space environment. The flight of *Voskhod 2* was a mystery to the West. No one knew at the time that the craft was a modified version of their one-person *Vostok,* and Soviet strategists were happy to let the West assume it was an entirely new vehicle. The airlock was an inflatable tunnel on the outside of the craft. Leonov entered the airlock tunnel, sealed the hatch to the spacecraft, and then bled off the pressure until he could open the outer door. Leonov's space walk was a daring experience, and nearly ended in disaster. He said the experience was exhilarating, like a "seagull with arms outstretched." For many years, Leonov would try to capture, in his paintings, the golden glow of Sunlight on his spacecraft. But when it was time to reenter the airlock, his pressure suit had ballooned in the vacuum. Leonov described it as being "like a big loaf of bread. Things do that in space." He was able to vent his suit through a valve and finally made it back into the airlock after ten minutes of struggle.

As the Soviet Moon program progressed, Leonov was seen as the best candidate to leave his bootprints on the lunar surface. He initially trained to circumnavigate the Moon before the U. S. could get their *Apollo* there, but problems with the modified *Zond* spacecraft (called the *L1,* essentially an adapted *Soyuz*) and

N-1 booster prevented the mission from happening in time. Soviet chief designer Sergei Korolev decided instead to try for a landing soon, and Leonov's cool head in *Voskhod 2* put him at the head of the line. In his excellent book, *Two Sides of the Moon*, the hard-working Leonov says the ultimate cancellation of the Soviet lunar program—in most part due to the N-1 failures—was "a devastating personal blow."

General Alexei Leonov (Photo courtesy and © of Tom Hunt.)

a test vehicle for future Mars missions. It continued to transmit data to Earth from the distance of Mars orbit, although Mars was not nearby.

The successes of Soviet orbiters and NASA's Lunar Orbiter program handed selenographers the data they needed to assemble detailed geological and topographic maps, narrowing down landing sites for human exploration. The robotic exploration of the Moon proceeded at breakneck speed. But on the human front, NASA's plans soon came to a tragic halt.

THE FIRST *APOLLO*

Before the clear evening of January 27, 1967, NASA's manned space program seemed blessed. Although many Western observers speculated that Soviet cosmonauts had been lost in launch or orbital accidents, no astronaut had lost his life in an American spacecraft. Through the sixteen flights of *Mercury* and *Gemini,* NASA flight engineers had accrued an impressive record of experience and triumph. There were failures and close calls, some quite dramatic, but none that caused the loss of a crewmember. NASA and the astronaut corps tried to continually remind the public that exploration is a dangerous prospect, and that loss of life is always a possibility. But the long march of successes lulled many into a false sense of safety.

Apollo 1 stood 200 feet above launch pad 34A at the Kennedy Space Center in the hot Sun of Florida, nested atop a Saturn 1B booster. Astronauts Gus Grissom, Edward White, and Roger Chaffee had been in their *Apollo* spacecraft for nearly five hours, rehearsing for the upcoming maiden flight of the *Apollo* CSM. No LM was cradled within the booster for this practice flight. The *Apollo* CSM was completely sealed, its interior pressurized with pure oxygen, just as it would be in space. Problems had hounded the crew throughout the grueling afternoon. Grissom's air system had developed a sour smell, delaying the mock countdown for over an hour. Various alarms continued to interrupt the test. The most critical issue was that of communications. The crew intermittently lost communications with the control room and launch blockhouse. At one point, Grissom quipped, "How are we going to get to the Moon if we can't talk between three buildings?"

Grissom—the leader of the team—pushed on. Problems were to be expected on the maiden voyage of a complex ship, especially one destined for the Moon.

At 6:31 pm EST, technicians in the clean room were preparing to "pull the plugs," severing the tethers and telemetry lines connecting the spacecraft to the gantry. They were stopped short by an ominous garbled message from the cockpit. The only clear word was "fire."

The crew of Apollo 1 *(left to right): Gus Grissom, Ed White, and Roger Chaffee. (NASA)*

Observers in the control room spotted a glow behind the window of the spacecraft hatch. Another communication came from the crew, this time from Roger Chaffee. "We've got a fire in the cockpit!" Chaffee sat on the far right couch and was tasked with keeping in communication with the ground in the event of emergency.

In the TV monitors, Ed White's arms were visible as he reached over his head to unbolt the ungainly hatch. Chaffee's next communication was urgent and heart-rending: "We've got a bad fire…We're burning up in here."

As flames glowed in the spacecraft windows, an explosion erupted from the side of the module, throwing pad leader Don Babbit to the floor. Momentarily disoriented and panicked, Babbit, mechanical technician James Gleaves, and systems technician L.D. Reece fled through the doorway, but all three turned back immediately to save the astronauts. They searched for gas masks and fire extinguishers, taking turns in the searing heat to unbolt the hatch. Grissom and White had been in the laborious process of opening the hatch—a process that would take 90 seconds under normal circumstances—when they lost consciousness. The crew fell silent within seventeen seconds, overcome by smoke inhalation. It took the desperate technicians more than five minutes to open the hatch, which was designed to be sealed from the inside.[5] Physicians Alan Harter and Fred Kelly were standing by in the blockhouse and rushed to the launch pad. They arrived less than fifteen minutes after the first alarm. The sight before them would change the course of history. Technicians managed to put out the intense fire and unbolted the outer and inner hatches. The normally white spacecraft and clean room stood scorched and blackened. Melted oxygen hoses hung from the spacecraft ceiling. Smoke still poured from inside the capsule, but the interior was now visible. Three American heroes had fallen.

The *Apollo* CSM brimmed with flammable materials. Velcro, affixed to the walls and floor, served to hold various items in a weightless environment. Nylon netting below the couches protected the interior from dropped equipment. Foam padding, present only during the test to protect fragile surfaces, provided more fuel for the fire. Within *Apollo's* pure oxygen environment, all these materials contributed to an opaque, acrid smoke.[6] Mercifully, the astronauts did not die of burns, but of asphyxia.

NASA convened a commission immediately. The agency was remarkably transparent in their work and studies. No one in official circles could complain that NASA was trying to cover up results. The board's findings pointed to the fire's flashpoint in some wires that may have come in contact with a metal surface. But the capsule had a host of problems, and—race or no race—it was past time to fix them. *Apollo 1* stood as a somber reminder that lunar exploration comes with inherent risks, and those risks will follow us into the future as humans establish permanent residence away from Earth.

Over the next 21 months, the commission's recommendations played out in a series of redesigns. Wherever possible, combustible materials were replaced with non-flammable ones. Engineers relocated areas containing non-metallic materials in such a way as to serve as firebreaks. They retooled oxygen-related systems to be more fire-resistant. Finally, the hatch was completely redesigned for easy ingress and egress. *Apollo* was pronounced good to go.

TRAGEDY IN THE SOVIET MOON PROGRAM

The Soviet space program soon became embroiled in its own tragedies. Just three months after the *Apollo 1* fire, Soviet flight engineers commenced the first flight of their new spacecraft, *Soyuz.* Although *Soyuz* had not been

5. This practice stemmed from the second *Mercury* flight, when Gus Grissom's hatch accidentally blew open in the ocean. The hatch was equipped with explosive bolts and opened from the outside. The ship sank and almost took Grissom with it. *Gemini* and *Apollo* capsules were outfitted with mechanically opened hatches to avoid such an incident, and the *Apollo* hatch opened inward to insure that the hatch would not fail in the vacuum of space.

6. All human-rated spacecraft today use a much safer mix of nitrogen and oxygen.

ISS017E005012

The Soyuz *spacecraft has an excellent track record for robust flexibility in many space operations. (Photo courtesy of NASA.)*

7. In fact, the *Soyuz* and the *Soyuz*-based *Progress* continue to be the backbone of the Soviet manned program, with regular flights to the International Space Station.

8. *Voskhod 1* was the first spacecraft to carry multiple people, beating the U. S. *Gemini* two-person craft into space by nearly a full year. It was a redesigned *Vostok*— the first Soviet human-rated craft— with so little room that cosmonauts could not wear pressure suits.

designed for lunar exploration, the Soviet lunar command module was based upon a similar design. Additionally, *Soyuz* was to be the workhorse of Soviet space activities for decades to come.[7] It was an important craft, and *Soyuz 1* was a pivotal mission. The Soviet team chose Vladimir Komarov to pilot the craft. He became the first cosmonaut to travel into space multiple times, after having flown *Voskhod 1*[8] in October of 1964.

The flight had problems from the start. One of two solar panels failed to deploy, depriving the craft of half its power. Then, the attitude controls of the craft began to degrade. Not only was Komarov left without automatic steering, but he was loosing some degree of manual control as well. The head of the program, Vassily Mishin, decided to cut the flight short, ordering Komarov down after his eighteenth orbit. Aboard *Soyuz 1,* Komarov executed a perfect reentry. After the blackout period that accompanies the fiery stages of descent, the cosmonaut reported that everything was normal. The first sign of trouble came when a pressure sensor designed to open the main parachute malfunctioned. Komarov deployed the emergency chute manually, but its lines fouled. Komarov's capsule plummeted to the farmlands below, exploding on impact. The loss of the beloved Soviet hero was as devastating to the Soviet Union as was the loss of the *Apollo 1* crew in the United States. And, just as *Apollo* had done overseas, the *Soyuz* disaster brought the Soviet Moon program to a full stop.

Soyuz was the first of a one-two punch taken by the Soviet program. The second was an even more serious disaster in relation to the outlook of the Moon program and involved their booster.

BEHEMOTH LAUNCHERS

Simply put, the biggest challenge to getting people to the Moon and returning them home again is the launch vehicle. It was true in the days of *Apollo,* and it is just as true for today's Constellation engineers. Transporting enough equipment for the round-trip journey requires a mind-boggling amount of power. Space-race engineers on both sides of the world scrambled to fashion a booster that would be up to the task. It wasn't easy. New, high-performance engines had to be developed and tested. Never-used alloys contributed to weight savings in the structures, and work began on the largest launch pads in history.

America's hopes rested upon the Saturn V. Its launch and flight strategies and powerful upper stages foreshadowed the future Ares boosters. With the *Apollo* CSM and escape tower on top, Saturn V towered some 110.6 meters (363 feet) into the Florida skies.

Five F-1 engines powered the first stage, generating 7.65 million pounds of thrust. By comparison, the space shuttle's three main engines generate about 1.2 million pounds of thrust. The first stage burned for 2 1/2 minutes.

The Saturn V is still the most powerful booster ever successfully flown.

By the time its fuel was spent, the vehicle was traveling at 6,000 miles per hour. The second stage, accelerated by five smaller engines, brought the vehicle to the outer fringes of the atmosphere at an altitude of 115 miles. At stage separation, the craft was coasting at over 15,000 mph. The third stage then kicked in, bringing the stack of Moon craft into Earth orbit. Its single engine

The Grand Plan from a Soviet Perspective

The Soviet scenario for lunar exploration was similar to the American plan in that all spacecraft involved in the mission were carried on a single booster. At the top of the N-1 booster, nestled beneath a safety shroud and escape tower, the four elements of the L3 complex were stacked one on top of the other. The base element was a fourth stage that would send the complex out of Earth orbit and on its way to the Moon. Above it, a fifth stage executed midcourse corrections and lunar orbit insertion, much like *Apollo's* CSM. Above the fifth stage, a

Two views of the Soviet LK manned Moon lander, on display at EuroDisney. Note the exiting cosmonaut for scale at left, and the down-facing window for landing (right). (Photos © by Michel Koivisto and courtesy of Karl Dodenhoff, myspacemusuem.com.)

lunar landing craft sat just below a *Soyuz*-heritage lunar orbiter with enlarged crew quarters and an engine to return a crew of two cosmonauts to Earth.

Unlike *Apollo*, the fifth stage would drop the lunar landing craft to within a mile or so of the surface. The lander was roughly spherical, with four landing legs similar to the American LM. The Soviet lander needed only one engine that would bring a single cosmonaut to the surface for up to two days. The Moon-walking cosmonaut's suit had the capacity to stay outside for only two hours or so. After collecting rocks, deploying instruments, and planting the Soviet flag, the cosmonaut's craft would lift off using the same engine, leaving behind only the landing legs. In lunar orbit, the cosmonaut would do a space walk to transfer himself and lunar samples back to the *Soyuz* return craft. As with *Apollo*, the lunar landing craft would be discarded before the return to Earth.

was designed to restart, enabling the stage to send the CSM and LM on a trajectory to the Moon. From there, the main engine of the CSM performed all the work of lunar orbit insertion, rendezvous and docking in lunar orbit, and return to Earth.

The Soviet approach to a Moon rocket was the secret—and massive—N-1. The launcher was only slightly shorter than the American Saturn V, measuring 5 meters less, but the conic first stage—known as the Block A—was far wider at the bottom to accommodate 30 engines. The cluster generated a thrust of 10 million pounds, far outpacing that of America's rocket. A series of 24 fixed engines ringed the outer edge, while six steerable engines formed a core group. A second stage utilized eight engines, while a third brought the L3 Moon craft into Earth orbit with four smaller engines.

The complexity of firing thirty engines in concert was a significant problem for the N-1. A complex safety system called KORD monitored engine performance and, should one engine fail or need to be shut down, the KORD shut down the corresponding engine on the other side of the booster. By 1968, the Soviet Moon program was over budget and behind schedule. The Americans had already launched two of their Saturn V's in unpiloted tests.

The first N-1 launch, on February 21, 1969, was an unmitigated disaster. Because of time and budget pressures, the Soviets opted to erect their new booster on the pad before they had static tested all first-stage engines together. Moments before liftoff at the Baikonur launch complex, the KORD system inexplicably shut down two healthy engines. Even with only 28 of its 30 engines burning, the N-1 leaped from the pad much more rapidly than the stately Saturn Vs. But one of the silenced engines burst into flame. The fire spread throughout the first stage, sending the mighty booster to an explosive crash 28 miles (45 km) from Baikonur.

While the vast N-1 launch complex was under final construction in 1968, the Soviets launched another pair of *Zond* spacecraft. *Zonds 5* and

The N-1 booster's first stage was designed to fire thirty engines in concert. The screen-like flaps were used to keep the rocket upright on the pad, and were left behind at launch. (Art © and courtesy of Nick Stevens.)

Rare photo of two N-1 boosters being readied for launch. (Photo courtesy of NASA/ Asif Siddiqi.)

6 carried biological payloads, and *Zond 5* hosted a mannequin in the pilot's seat. To American analysts, these flights were alarming, as they followed a "free return" trajectory—a figure eight path—looping around the Moon and returning home without having to make any maneuvers. If they could do it with mannequins, were they ready to do it with cosmonauts?

Added to U. S. worries was the news from the Pentagon. The months of launch preparations for the N-1 had not gone unnoticed in the halls of the Defense department. Spy satellites had imaged the vast launch complex at Baikonur, and it was clear the Soviets had a new booster that was big enough to go to the Moon carrying a payload of cosmonauts and equipment. NASA had successfully carried out the first manned flight of the newly designed *Apollo* CSM in Earth orbit. *Apollo 7* launched aboard a Saturn 1B with seasoned *Mercury* astronaut Wally Shirra, Don Eisele, and Walt Cunningham. The Saturn 1B was a smaller sibling to the mighty Saturn V Moon rocket and was not strong enough to carry an *Apollo* to the Moon. But it was fine for lofting the *Apollo* CMS into an eleven-day spin around Earth. Plans for the next flight called for another Earth-orbit test involving the *Apollo* and the LM. But development of the LM was behind schedule. NASA made a remarkably bold decision: send the next crew into orbit around the Moon instead of waiting for a LM to test near Earth.

THE FIRST TRUE MOON MISSIONS

Apollo 8 was daring in many ways. It was the first manned flight of the mighty Saturn V. It was the first flight to escape Earth's gravity. And it was the first flight into lunar orbit. Unlike the *Zond* free-return flights, the astronauts would brake into orbit around the Moon. If anything went wrong with the relatively untried service module engine, the crew would be stranded. But the highly trained pilots knew the risks, and they knew their spacecraft. They had confidence in the thousands of engineers and mission control personnel who would send them across the void.

Veteran astronaut Frank Borman commanded *Apollo 8*. The CSM pilot was James Lovell, another veteran. Rounding out the crew, rookie Bill Anders was designated as LM pilot. Although *Apollo 8* carried no LM, the position was a critical one for this complex, trailblazing flight. The astronauts coasted across the translunar gap in the course of three days. *Apollo 8* passed behind the Moon before its main engine fired to place the craft into orbit. The flawless burn set up a ten-orbit visit above the Moon. On Christmas Eve, the crew described the forbidding lunar surface and read from the book of Genesis. The televised moment was the most watched broadcast up to that time in history.

Apollo 8 made it home on December 27. In the midst of the Vietnam War and civil unrest in many places, a remarkable photograph appeared in newspapers and magazines across the world, an image of a glistening blue Earth floating against a velvet-black backdrop of space, the desolate lunar

Astronaut William Anders snapped this historic image of the first Earthrise witnessed by humans. Although the photo is often printed on its side, in this orientation north is at top. (Photo courtesy of NASA/JSC.)

landscape spread below. The image, snapped by Bill Anders, became the visual call-to-arms of the environmental movement and a reminder that in all the turmoil and strife, there is only one Earth, and it is a place to be cherished. *Time Magazine* voted the *Apollo 8* crew as their "Men of the Year" for 1968. As one well-wisher's telegram—sent to Frank Borman—put it, "You saved 1968."

The next step in fulfilling President Kennedy's national goal was a test of the LM. The ungainly craft was unique: it was the only manned spacecraft ever designed to fly exclusively in space. Echoes of the LM are seen in designs for the Constellation suite of lunar craft, all vehicles designed to operate in a vacuum. The LM's fuel and oxygen tanks hung at odd angles on the outside of the main cabin, while the craft bristled with communications antennae and radar dishes. The LM needed no streamlining, as it operated in a vacuum. Patches of black and white paint butted up against gold foil, all designed to keep various systems at just the right temperature in the extremes of space.

The maiden flight of the lunar module fell to the crew of *Apollo 9*. It was March of 1969, and Kennedy's deadline loomed. The Soviets were also busy with activities mysterious enough to keep the U. S. Defense Department and

The LM coasts above Earth in its maiden flight during Apollo 9.

NASA guessing. A successful landing depended on a working LM, and it was time to see if the talented engineers at Grumman Aerospace had come through.

The LM turned out to be one of the most successful engineering feats of NASA's human spaceflight program. The spindly two-stage beast sprouted four landing legs for the lunar surface, advanced navigation systems, and complex rendezvous radar, all of which had to work flawlessly, as lunar explorers' lives depended upon it. And the LM worked every time.

Apollo 9's crew docked and undocked the *Apollo* and LM, transferred between spacecraft, tested the *Apollo* spacesuit (the first to have an independent life support system), and conducted several space walks. While Dave Scott piloted the command module, James McDivitt and Rusty Scheickart flew the LM 111 miles (133 km) away, then cast off the descent stage and used the ascent stage to return to *Apollo.* The spacecraft and new suits worked well. All was ready for the next stage: a dress rehearsal in lunar orbit.

Apollo 10 flew a similar path to that of *Apollo 8,* but this time the command and service modules were linked to a lunar module, fully outfitted for a powered descent to the Moon. Commander Tom Stafford and LM pilot Gene Cernan flew within 8.4 miles (15.6 km) of the lunar surface. Overhead, John Young readied the CSM in case he needed to swoop down to rescue them. As the LM coasted over the rugged lunar landscape, Cernan and Stafford commanded the ascent stage to fire, a rehearsal for a lunar liftoff.

A Soviet attempt to launch the titanic N-1, this one under cover of darkness. (Photo courtesy of NASA.)

For a few moments, the craft spun wildly, but the pilots quickly regained control and returned safely to the CSM. Had this been a real launch from the lunar surface, the crew would have been lost. Engineers tracked the problem to a series of minor failures that, in concert, would have contributed to a space catastrophe. The problems were easily fixed.

After years of testing and retesting, starts and false-starts, sacrifices and engineering marvels, NASA stood ready to send the first humans to the surface of the Moon. Launch was scheduled for July of 1969.

While the Americans tested their Moon exploration vehicles in Earth and lunar orbit, the Soviets struggled to get N-1 into the air. Just three weeks before the first American attempt to land people on the Moon, the second N-1 fired thirty engines and began to rise off its pad. But less than one second before liftoff, a fragment of metal lodged in one of the fuel pumps. The KORD system dutifully shut off the engine, along with its counterpart on the other side of the booster. The N-1 lifted into the sky above Baikonur some 200 meters before the lower stage disintegrated. The entire launch area was destroyed.

Two weeks later, as the crew of *Apollo 11* prepared for the first flight to the lunar surface, the Soviets attempted a scientific and political coup. Their goal: return samples from the Moon robotically. *Luna 15* was launched on July 13, 1969. The craft made it into low orbit around the Moon just a day before the arrival of *Apollo 11*. NASA observers kept close watch as the spacecraft made 52 orbits of the Moon. On July 20, *Apollo 11*'s LM, called the *Eagle*, touched down in the Sea of Tranquility, but NASA observers were still nervous. Official requests to the Soviets for clarification about their mystery satellite went unanswered.

Michael Collins became the most solitary person in history when the LM separated, leaving him in orbit in his CSM. While astronauts explored the Moon, Apollo *CSMs stood watch, ready to descend to low altitude if the LM suffered a malfunction. (Photo courtesy of NASA.)*

As *Luna 15* circled, so did Mike Collins in the *Apollo* CSM called *Columbia.* Collins continued to do science from orbit while Armstrong and Aldrin descended to the lunar plains. The descent was somewhat harrowing, especially to ground crews. The flight computer aboard the LM sporadically chimed an alarm that no one had trained for. The computer's message, "1202," meant that it was overloaded and threatening to recycle back to an earlier state. This would have been disastrous for the crew, as they were dropping toward the surface. But *Apollo 11*'s seasoned pilots did not panic, and in the final moments, with fuel running low, manually piloted the spindly spacecraft to a gentle landing.

The day after the LM touched down, *Luna 15* engaged its landing sequence, but crashed into the Sea of Crises on July 21. The destruction of *Luna 15* sounded the death knell of the Moon race. Neil Armstrong and Buzz Aldrin walked across the finish line on the dusty lunar plains while the world watched. The landing made the covers of every magazine from *Life* to *Paris Match.*

When Armstrong took his "one giant leap for mankind," his thoughts were focused on the gravity of the historic moment.[9] But those thoughts were soon interrupted by NASA's unforgiving timeline. History's first Moonwalk lasted 2 hours and 31 minutes. NASA played the agenda conservatively. Armstrong and Aldrin took panoramic photography, scooped soil samples, bagged precious moon rocks, and took several core samples. They also set up the first ALSEP (*Apollo* Lunar Science Experiment Package), a suite of instruments designed to continue surface science after *Apollo 11* departed.

The historic flight of *Apollo 11* was followed, four months later, by *Apollo 12.* The Saturn V rose through an electrically charged Florida sky, its long plume of exhaust becoming a 6,000-foot-tall lightning rod. Lightning knocked out most of the major electrical systems, and the astronauts had to reboot several electrical systems while under the increasing G-forces of ascent, making for a nerve-wracking launch. In orbit, extensive tests assured the crew and flight controllers that all systems were normal and undamaged.

Apollo 12's LM may have landed in the Sea of Storms, but it was smooth sailing from its arrival. Commander Pete Conrad and LM pilot Alan Bean were relaxed in the lunar environment, wise-cracking and joking with each

9. Armstrong meant for his words to be, "That's one small step for a man, one giant leap for mankind." Whether he remembered the "a" remains a point of debate.

other and mission control a quarter of a million miles distant. The 5'6″ Conrad was the shortest member of NASA's astronaut corps. Conrad's first words on the lunar surface were, "That may have been one small step for Neil, but it's a long one for me."

The lunar outing had begun well, right from the landing. Months earlier, *Apollo 11* had set down nearly 5 miles from its target point, and flight engineers were banking on *Apollo 12* to demonstrate that a more accurate landing was possible. To that end, mission planners challenged the crew to hit a landing site within walking distance of an earlier lunar explorer: *Surveyor 3.* Conrad and Bean came through with flying colors, landing within 156 m (512 ft) of the robot pioneer. "The improved techniques we

The first photo taken by a human explorer from the surface of the Moon. Neil Armstrong snapped the shot almost immediately after stepping onto the surface in case anything went wrong and the crew had to leave in a hurry. (Photo courtesy of NASA.)

developed for our flight were then used and refined in all future flights," says Alan Bean. "The next flight had to do things that were different. Each flight in a space program is new. To the outsiders, it seems same old thing, but when you look at it, we're always developing things new and increasing the capability of America to explore space." The pinpoint touchdown confirmed a capability that would be critical for future missions. Later *Apollo* flights were targeted for narrow landing targets between mountains or along the edge of valleys.

The crew of *Apollo 12* carried out two EVAs (extra-vehicular activities, or Moonwalks), totaling just under eight hours, nearly quadrupling Armstrong's and Aldrin's surface time. On the first EVA, the astronaut duo gathered samples fairly near the LM, pounded core tubes into the remarkably dense lunar dirt, and set up various experiments.

After a seven-hour sleep period, the second EVA featured the long-anticipated trip to *Surveyor.* As Bean and Conrad approached, they could see the silent sentinel resting in the ancient soil, undisturbed for 2 1/2 years, save for a thin coating of dust deposited by the LM's engines at arrival. *Surveyor*'s dead camera stared out across the rolling plains, its soil scoop poised above the ground as if awaiting a final command to dig. The astronauts removed several items from *Surveyor,* including its camera and soil scoop. Laboratory technicians would later discover freeze-dried bacteria within the foam insulation of the camera assembly. The *Streptococcus mitis* was subsequently cultured and identified at the Center for Disease Control

Al Bean poses next to the Surveyor 3 on the gentle slope of a crater rim. The LM is close by.

in Atlanta. The bacteria had survived all vacuum chamber and heat tests on Earth, three years of severe lunar temperature swings of 300° F, and a complete lack of water and air.

Apollo 12's return to Earth marked a nearly flawless mission with a high scientific return. Mission planners promised much more to come. Although *Apollo 11* and *Apollo 12* provided the first human survey of the lunar environment, ensuing missions became increasingly complex. *Apollo 13* was to be the final of the first survey missions. After its exploration of the Fra Mauro highlands, the new "J" missions would fly. These missions had LMs with extended capacity for fuel and oxygen. They carried a lunar rover that would enable astronauts to travel not yards but miles. The *Apollo* program was now in full gear, and the future looked bright.

However, by April of 1970, Americans had lost interest in the Moon explorations. It had only been nine months since that first historic landing,

Crises in the Cosmos: *Apollo 12*

On November 19, 1969, Alan Bean became the fourth human to walk on the Moon. Getting there was half the challenge. Astronaut Bean shares his observations on what it takes to get to the Moon, to train for the unpredictable, and to survive the overwhelming experience of landing on another world.

We all came from a background of flying high performance airplanes. That is the closest analog to flying spacecraft. The LM is just a more complex flying machine. First, it flew like an airplane, and then for landing it flew like a helicopter. We all had to learn to fly helicopters to develop those skills. We developed simulators, and our technology for simulations has now spread all over the world. We had to become proficient in simulators, because you can't learn on the job with fuel that's going to run out about a minute after you are supposed to have landed. That's why we had the success we had. Neil

Armstrong had to take over and find a place that was a little smoother to land, and he was able to do that. Without the simulators and training and self-confidence that gave him, he wouldn't have been able to do that. I remember Pete [Pete Conrad, *Apollo 12* LM pilot] saying that he had to use all the skills that he had developed over the years of flying airplanes, making carrier landings, and simulators to make the landing that he made on *Apollo 12*. These are exactly the same kind of skills that crews returning to the Moon are going to have to learn— Chinese, American, what have you—they're going to have to practice it over and over again in simulators.

[When the lunar surface first came into view during descent] as soon as I noticed I was fairly excited and kind of scared, I said to myself, "This is not conducive to good performance." As an astronaut and test pilot, you learn to adapt. So I said, "If I look inside and pay attention to the computers and my displays, then I won't be excited by looking out the window." When I looked inside, it looked just like the simulator. I didn't want to miss the trip, but I waited until I settled down and my heart rate dropped before I looked out again. It was still exciting, but not quite as much. I was adapted. Whenever I felt myself feeling distracted or paying too much attention to what was outside, then I'd look back in and it felt just like the hundreds of hours we'd spent in the simulator.

During Bean's second EVA, he experienced a potentially life-threatening crisis.

I was running along the surface, and all of a sudden my ears popped, which immediately tells you that you are either gaining or losing pressure. That's a big deal, because you lose very much and you're out of there; if you lose it all, you can be dead in just a few seconds. I stopped and looked at my cuff, at my pressure gauge. I didn't say a word, but Pete knew something was going on. He said, "What's up, Al?" My gauge said everything was right: 3.7 psi. I stood there looking at it, and I asked Houston to take a look at it. They couldn't see any change. So we just went about our business. We found out later that as I ran along, my stomach and the suit exhaust valve touched, and my stomach kept the O_2 [oxygen] from going back to the PLSS (the Portable Life Support System backpack), and that actually raised the pressure in my suit. The engineers put a gasket around the valve so that it couldn't be closed off in future flights. That's test flying; that's what spaceflight is all about. One test flight is worth a thousand expert opinions.

Alan Bean became the fourth man to walk on the Moon.

and the second had gone like clockwork, so much so that the excitement seemed to be draining away. The collective American mind turned to the costly war in Vietnam, the social upheavals at home, and a mounting economic recession. When the crew of *Apollo 13* signed off of their evening broadcast on the way to the Moon, few were watching. Within the span of less than a year, the Moon program had gone from the greatest adventure in history to old hat.

That was about to change. *Apollo 13* had been coasting through the void for two days. The crew aboard the CSM called *Odyssey* were nearly 200,000 miles (322,000 km) from Earth. Periodically, the crew stirred the oxygen tanks to keep super-chilled liquid oxygen "slush" from settling out. The oxygen tanks served not only to charge the atmosphere in the *Apollo* and LM but also as part of the fuel cell power plants. Oxygen flowed through a matrix into hydrogen. The two elements combined to form the crew's drinking water and electricity. At one point as the "stir" command was sent to the service module, the crew heard a loud bang.

Alarms blared. Mission commander Jim Lovell saw gases spurting from the side of the CSM. The ship began to turn in space, and Jack Swigert spotted a warning light that indicated a precipitous power drop. Jim Lovell then relayed the famous words to Mission Control, "Houston, we've had a problem." Fred Haise, who was down in the LM preparing for the next day's descent, rushed back through the docking tunnel to see what was wrong. As he passed the connecting ring between the two spacecraft, he heard it popping and groaning as the two craft rocked back and forth. Warning lights for many unrelated systems continued to glow. Had *Apollo 13* been struck by a meteor? Had something gone wrong with the fuel?

The CSM had only minutes of power left. Without it, the crew had no way home. All navigation data was in the dying CSM, but the LM *Aquarius* had not only navigation equipment but oxygen and fuel. The LM became the first space lifeboat. The crew feverishly transferred data from the CSM into the LM's computers. Then, they shut down the command module, putting it into hibernation mode. However, to save fuel, temperatures in the LM were allowed to fall, and carbon dioxide levels soared as engineers on the ground tried to devise fixes to keep the crew alive. The LM's main engine had to be used to get the *Apollo* out of lunar orbit and back to Earth. Prayer vigils sprang up all over the world as the people of Earth became one in their hopes for the three men a quarter of a million miles from home.

In the last stages of the mission, the crew had to do something that years of training told them not to: instead of letting go of the LM, they would jettison the service module. The LM still had to get them closer to Earth, and there was not enough oxygen or battery power in the command module to power it up yet. The crew was cold, exhausted, and dehydrated from water rationing, and Rusty Swigert knew that if they followed normal procedures, they might make a mistake that could kill them all. He put a piece of tape over the button marked LM JETTISON so that the lifeboat would remain intact until the last minute.

With the LM still attached, the explosive bolts holding the Service Module fired. Now, *Apollo* was in a configuration never tested for: a lone command module attached to the LM. The flight engineers and crew hoped to get a look at the crippled service module to search for damage, but the trip had been so harrowing that no extreme measures would be taken. As the giant cylinder floated away from the *Odyssey,* Lovell turned the LM and command module. He radioed that, "There's one whole side of the spacecraft missing!" The explosion of the fuel cell's oxygen tank had blown off the entire covering of nearly a quadrant of the craft.

The crew awakened their command module, sealed the hatch that led to the LM, and bade farewell to the good ship *Aquarius.* Hours later, they were safe aboard the aircraft carrier Iwo Jima. The mission was a failure in terms of lunar exploration, but it was an unprecedented success in long-distance engineering and the triumph of the human spirit.

After the harrowing flight, engineers traced the failure to a change in the electrical system. *Apollo*'s wiring had been upgraded from 28 volts to 65 volts

Lovell's view of the service module, with an entire side blown off, chilled him as much as the low cabin temperatures of the crippled Apollo 13.

so that it could accept higher-voltage ground tests. Designers had missed one upgrade: the thermostat inside the liquid oxygen tank. A ground test of *Apollo 13* had damaged the thermostat and fried some interior wires. When the fan was turned on to stir the tank, the wires arced and ignited the liquid oxygen.

While the *Apollo* service module was being redesigned for the next flight, the question continued to haunt the halls of the Johnson Spaceflight Center: where were the Russians? It was clear, from orbital imagery, that they continued to work on the launch complex at Baikonur.

The Soviet Union launched a reply to NASA's manned *Apollo* program in September of 1970, when it succeeded in getting *Luna 16* down to the lunar surface and back to Earth again. While the craft returned only 101 grams of lunar material (*Apollo 11* and *12* had brought back a total of 22 kg), it was a spectacular robotic mission. *Luna 16* would be followed by two other sample return missions. *Luna 16*'s automation would later be echoed in some of NASA's Constellation designs for robotic and human-precursor missions.

In November of 1970, the Soviets racked up another space "first" by deploying a lunar roving vehicle aboard *Luna 17*. The *Lunakhod 1* survived for eleven months, drove 10,500 meters (6 miles), and returned over 20,000 images.

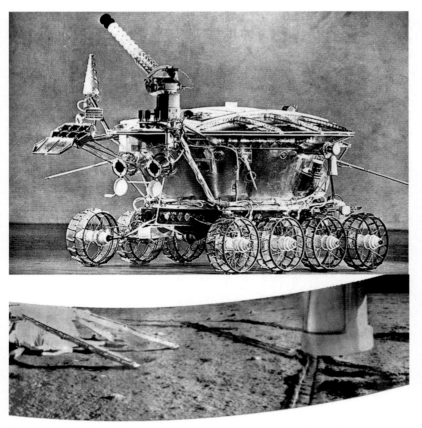

The Soviet Lunokhod 1 *(above) returned curving panoramas (below) from the Sea of Rains. (Top photo courtesy of NASA; bottom panorama courtesy of Don Mitchell.)*

The redesigned *Apollo 14* made it to *Apollo 13*'s target on February 5, 1971. The crew clocked 33 hours on the lunar surface. Astronauts Alan Shephard, Jr., and Ed Mitchell spent 9 hours and 22 minutes in two EVAs, collecting the first lunar samples of highland materials, the bright material making up the mountains of the Moon. In orbit overhead, Stu Roosa deployed a satellite, the largest payload ever left in lunar orbit to that time. *Apollo 14* paved the way for the far more capable and extensive "J" missions of the next three *Apollos*. But the Soviets weren't finished in their quest for the Moon.

On June 27, 1971, just a month before the scheduled launch of *Apollo 15*, the third N-1 booster blasted into the skies over Baikonur. The massive rocket passed the 250 meters altitude mark when the guidance system failed, sending the craft into a fatal spin. The second and third stages separated. The first stage explosion left a 30 meter (100 foot) crater.

N-1's redesigned first stage debuted in November of the following year, but vibrations tore the craft apart after 107 seconds. Although two more boosters were fabricated, the program was quietly canceled in 1974. Sadly, the last flight of the N-1 in 1971 signaled the death rattle of the Soviet manned Moon program. The Soviet Union did not acknowledge the existence of its program until 1989.

THE "J" MISSIONS: *APOLLO*'s Crescendo

NASA's *Apollo* program came to full fruition with the flights of *Apollo 15, 16,* and *17*. Each carried a lunar rover and far more equipment than earlier flights. Improved suits enabled astronauts to stay outside the LM for longer periods, and improved flying techniques allowed crews to set down in more dangerous—and geologically significant—terrain. The first of the "J" missions, *Apollo 15* threaded its way through rugged terrain to make landfall in the Apennine Mountains, landing between a 15,000 foot peak called Hadley Delta and the abyssal canyon Hadley Rille.

In addition to the advanced hardware, there was another difference between this mission and previous excursions. Dave Scott and Jim Irwin had trained with geologists more extensively than any other crew. During three EVA's, the two men spent 18 1/2 hours outside the LM and traversed 17.25 miles (20.7 km) thanks to their rover. The rover ride was far more exciting than anyone had expected. Although the Moon buggy had a top speed of about 8 mph (12 km/h), the Moon's 1/6 gravity couldn't seem to keep the four wheels on the ground. Every time the astronauts hit a bump or hollow, the rover left the surface.

From the slopes of Hadley Delta, Irwin and Scott beheld the most spectacular *Apollo* landing site yet. The gunpowder-gray mountains appeared snowy against the

Jim Irwin works beside Apollo 15's lunar rover, perched on the edge of Hadley Rille. The soft lunar landscape is deceiving. Hadley is 2/3 as deep as the Grand Canyon.

black sky. Steep slopes curved down to the cratered plains, where the chasm of Hadley Rille cut a black serpent across the rolling gray hills.

The astronauts drilled deep core samples, deployed advanced instrumentation, and made observations in the field that only humans could have done. On the second EVA, Jim Irwin found what came to be known as the Genesis rock, an anorthosite rock that represents the early crust of the Moon. By various methods, the rock has been dated at 4 billion years old, nearly twice as old as many rocks from the Moon's "repaved" volcanic maria regions seen by *Apollo 11* and *Apollo 12*.

For a final encore, Dave Scott parked the lunar rover in a position that enabled audiences on Earth to watch—via the rover's remotely controlled TV camera—the launch of the LM from the surface of the Moon. With the quarter-of-a-million-mile delay in communication, camera operator Ed Fendell had to begin panning the camera upwards a full three seconds before the actual launch. As *Apollo 15* lifted off of Hadley Base, viewers on Earth watched the initial splatter of foil and debris at launch and the smooth liftoff of the LM's ascent stage as it carried two tired men back into lunar orbit.

Apollo 16 and *Apollo 17* continued to demonstrate the complexities of lunar exploration and the value of human presence in lunar science. *Apollo 16* landed in the Descartes highland region, while *Apollo 17* surveyed a dark

This dramatic view of the distant LM demonstrates the capability brought to missions by the lunar Rover. (Photo courtesy of NASA.)

valley nestled between the Taurus Mountains and the crater Littrow. *Apollo 17* had the distinction of carrying the first true scientist to the Moon, Harrison "Jack" Schmidt.

The *Apollo 17* site lay in a valley 1½ times as deep as Arizona's Grand Canyon. Scientists suspected that volcanism had invaded the area, perhaps depositing the dark material on which the sixth lunar lander settled. Geologists also hoped to get samples of even more ancient lunar crust. Much of the hemisphere of the Moon visible from Earth has been affected by the Imbrium Basin, a great scar that forms the left eye of the Man-on-the-Moon. The impact that gouged out Imbrium cast molten rock and debris across much of the Moon's near side, leaving craters and uplifts. Taurus-Littrow, it was hoped, was far enough away from the site that older material survived there. The 109 kg (240 lbs) of samples brought back included a small gray stone 4.5 billion years old, the oldest lunar sample yet found. Perhaps more importantly, the crew sampled debris from a landslide that researchers believe was triggered by the impact of the titanic Tycho crater. Analysis indicates that the sample had awaited Jack Schmidt's sample scoop for 109 million years, giving scientists a good estimate of the age of Tycho.[10]

Apollo 17 brought a dramatic end to NASA's productive and historic Moon missions. The mission commander, Gene Cernan, was the last person to leave the Moon. As he prepared to step up the ladder for the last time, the astronaut felt the full weight of the historic moment. As his words were broadcast to the entire world, the last *Apollo* Moonwalker said, "As I take man's last steps from the surface, back home for some time to come—but we believe not too long into the future, I believe history will record that America's challenge of today has forged man's destiny of tomorrow. As we leave the Moon at Taurus-Littrow, we leave as we came, and God willing, as we shall return, with peace and hope for all mankind."

With that, the last lunar explorers of the twentieth century rocketed into the Sun-drenched nighttime sky above the mountains of the Moon.

---

10. Recent research at the Southwest Research Institute reinforces this estimate. Computer models indicate that two asteroids collided some 160 million years ago. Debris from the collision, scientists contend, caused the impact at the end of the Mesozoic Age which wiped out the dinosaurs, as well as the impact that created the magnificent Tycho Crater.

Panoramas of the Apollo *landing sites. From top to bottom:* Apollo 11, *Sea of Tranquility;* Apollo 12, *Surveyor Crater, Ocean of Storms;* Apollo 14, *Fra Mauro;* Apollo 15, *Hadley Rille and Appenine Mountains;* Apollo 16, *Descartes Highlands;* Apollo 17, *Taurus-Littrow valley. Compare the smooth Maria plains in earlier missions to the rugged highlands of the later "J" explorations. (Photos courtesy of the Lunar and Planetary Institute/Nasa.)*

Chapter Two

Getting There the Second Time Around

The first launch of the mighty Ares V. (art by author)

When the first Ares V thundered into the Florida skies in 2018, the booster trailed not only a column of incandescent smoke but also a rich heritage spanning half a century. From the simple Scouts to the complex shuttles, from the diminutive Redstones to the mighty Saturn Vs, Ares topped a vast family tree whose roots stretched back to ancient

M. Carroll, *The Seventh Landing*, DOI 10.1007/978-0-387-93881-3_2,
© Springer Science+Business Media, LLC 2009

Chinese fire arrows a millennium before. Thousands of engineers and technicians labored to transform blueprints into boosters. And after years of development, static firings, wind tunnel tests, and computer simulations, it was finally time to set sail on the translunar sea.

Engineers and strategists have visions of hardware twinkling in their eyes. They dream of foil-encrusted cargo vessels settling upon spindly legs into billowing lunar dust. They revel in thoughts of glistening spacecraft carrying crews across the void, Moonward after a forty-year exploration drought. They pour over blueprints of bigger and better boosters, of faster ships steered by more powerful computers and advanced technologies. They are the dreamers, and realities begin with dreams. But behind those visions, there must be substance. There must be reason. And so the questions come: What happened in those years since the last *Apollo?* Why haven't we gone back to the Moon? *Should* we go back? Don't we know enough about this cold, dead world? Are there reasons to return beyond national pride, keeping up with the international Joneses, or a few dusty rocks? And should we instead be looking farther afield…to places such as Mars?

REASONS TO RETURN

To NASA Goddard's Chief Scientist Jim Garvin, "Space matters. We live in space, after all. Yes it's inspirational (look at the Hubble), but look at where space affects us. It's given us an understanding of our own planetary climate system and the dangers from space. Going to the Moon enables us to build a platform for technology that is transformational to our society, and that's not just Teflon. One example of those transformational things is the information technology for high reliability in space systems that has been applied to many aspects of our society."

Those applications have had four decades to sink in. No humans have returned to terra luna in that time. The reasons are a complex stew of finances, technological challenges, and political will. But across those years, technology has matured, seasoned by advances in computers, robotics, and materials. Many analysts suggest that the time is right to build a permanent human presence on the Moon. The scientists want to go back. *Apollo* left us with intriguing puzzles and mysteries, and it also left us with the understanding that knowledge of the Moon gives us insights into Earth. But reasons for a return go far deeper.

Technological Reasons

"We're still riding the wave of innovation that came out of the *Apollo* program," says Constellation Manager Jeff Hanley. "It wasn't so much the widgets that got built, but rather the precision, the way of going about building widgets of high precision and high reliability. That was the true benefit of the *Apollo* program. It infused industry with ways of doing business and standards of building things that were unparalleled at the time, because they had to be incredibly high in reliability, incredibly simple, incredibly low mass. That drove miniaturization." Hanley joins a widespread chorus of voices declaring the benefits of space technology to the general population. And while some analysts argue that certain areas of American technology are stagnating today, many believe that a vibrant lunar program will drive technology in new directions that will benefit the world's population and economy.

Constellation EVA Systems Project Manager Glen Lutz points out that, "The brakes in my car are better today because of the heat rejection problems NASA had to solve for reentry into Earth's atmosphere." Lutz believes the benefits to society are seldom seen ahead of time, but the applications are numerous. "For example, why spend this money on spacesuits? The resulting technology has given us cooling for medical procedures, techniques for radiation treatment, miniaturization of components for health industry, materials research, and the list goes on."

Added to that list are the commercial aspects of a return to the Moon. In 1991, commercial and governmental space spending was at $11.5 billion; by 2007 it had reached $251 billion.[11] Companies such as Google and Virgin Galactic are tapping into a new arena of private exploration and space tourism. The infrastructure built by NASA, ESA, and other spacefaring communities will enable commercial ventures only hinted at today.

Cultural Reasons

"Science and space exploration will drive humanity's search for knowledge in the coming generations, and we must recognize there are only so many things we can learn here on Earth, and give NASA the tools to explore the rest." These observations were not made by a scientist or space strategist but by Tom DeLay, then house majority leader, in a 2005 press conference. DeLay's comments came on the heels of a congressional act, signed into law, assuring funding for NASA. The NASA Authorization Act of 2005 (HR 3070) received overwhelming support from both sides of the aisle and found form in NASA's Vision for Space Exploration. The congressional act states, in part, that Congress must have a "clear policy and funding provisions to insure that NASA remains a multi-mission agency with robust R & D activities in science, aeronautics, and human space flight." The act also called for "support for the goal of human space exploration beyond low Earth orbit and guidelines to insure it is properly paid for and not funded at the expense of other important NASA programs." The bill was approved 383-15.

Does space matter to the American people, as Jim Garvin believes? A 2004 Gallup Poll showed that 68% of Americans supported the Vision for Space Exploration. By 2005, that approval number had jumped to 77%.

Gene Cernan has often remarked that the last Moon flight, his *Apollo 17*, was "the end of the beginning" of lunar exploration. NASA's vision for a return to the Moon is an attempt to put truth to his remark. But to get there with a new generation of explorers and equipment, an advanced series of

11. Reuters, Wednesday, April 9, 2008.

***Apollo 16's* John Young: Saving Earth through Lunar Exploration**

Captain John Young is one of the most experienced space travelers in history. He flew a total of six missions in the Gemini, Apollo command codule, lunar module, and space shuttle vehicles, and spent three days on the lunar surface in the Descartes Highlands with astronaut Charlie Duke. Recently, Young has focused public attention on the migration of the human species into space, and how application of the technology needed for such migration will assure its survival.

We're going back to the Moon not because we want to but because we must. It's to save civilization. If you look at the geologic record, single-planet species don't last. Take a look at the 300-mile diameter crater at the end of the Permian, 250 million years ago. [The asteroid that created that crater] wiped out everything on the planet, 90% of the species. There's nothing we can do to handle something that's going to make a 300-mile-wide crater, and there are plenty of asteroids and comets out there that can do it.

Going back to the Moon is very practical for the long haul of civilization. You industrialize the Moon. You're able to live and work up there. You're able to terraform (change an environment into an earth-like one). You have the kinds of things you need to protect people if bad things happen on planet Earth. I think going back to the Moon is really the key to our future. Just having a moon makes it possible for us to survive; once we industrialize the Moon, develop alternate energy sources, and generate solar power and ship it back to Earth we'll totally change the way people live on this planet. If you just look at the fossil fuels we'll be using when China and India [become completely industrialized] we'll be using so much that we're not going to make it. At the rate we're going, we're not going to last. Something has to change. [A return to the Moon] will give us the technology we need to control our own destiny.

The Moon is also the key to Mars. Once you learn to live and work on the Moon you can handle stuff on Mars. Now, Mars is going to be a little different because the dust floats. We have to learn how to deal with it on the Moon before we go to Mars. Having an airlock [on the Altair and rover] will help. Maybe you have an inner place where you clean up—a pre-airlock—or a place outside where you clean up before you even get into the airlock. Another possibility is an outer suit that you take off. We've been working on it. At this point, dust is in a lower category, but it would sure wipe you out. But I think going back to the Moon is the key to preserving civilization on this planet. The more we can do to industrialize the Moon, to learn to live and work up there, is really the key to our future.

crewed orbiters and lunar transports must be produced. Carrying the load will be the next generation of boosters, christened Ares.

ARES: THE NEW WAY UP

Instituting a new family of launch vehicles is a daunting task, and one that designers do not take lightly. The logical question asked in the beginning was, why not simply upgrade the expendable boosters we already have? To answer the question, NASA and independent study groups considered three areas: performance (necessary lift capability), risk (comparative reliability and track record of various existing systems), and cost of all approaches and systems.

The new boosters must enable *Orion* to take the place of the space shuttle. The shuttle is a powerful machine. To match its role, the *Orion* spacecraft will need to carry substantial human and cargo payloads into low Earth orbit. *Orion* will also be tasked with getting crews to the vicinity of the Moon. Ares Earth-orbit capacity must surpass 20 metric tons to orbit. It must also be able to transport 23.3 metric tons into a translunar orbit, a path that leads out of Earth's gravity and ends at the Moon.

These were the requirements. The next step was to see if any available launch systems could be modified to fill the bill, which involved scrutiny of commonly used systems such as the space shuttle's main engines (SSMEs) as well as evaluation of what are known as the evolved expendable launch vehicles (EELVs). Since the SSMEs were in relatively constant use, the powerful engines seemed a good bet for use in the next generation boosters. But other boosters were in contention with good track records and hardware that was available. Both the Delta IV and Atlas V, current workhorses of the U. S. space program, were in the running for adaptation to Ares.

Studies showed that both the Delta IV and the Atlas V have insufficient power to boost the large payloads called for in a Moon mission. Could they be safely modified, not only to carry humans but also to carry the extra weight into space? Engineers determined that a new upper stage would be required with high performance engines, but even this would not be sufficient. Planners then looked to strap-on boosters, smaller versions of the shuttle's solid rocket boosters that strap to the side of its external tank. The problem is that such solid fuel strap-ons lower the safety of the system and add complexity.

In fact, safety became the major concern. The most powerful EELVs were never designed to carry humans. In a speech to the Space Transportation Association, NASA administrator Mike Griffin said, "Significant upgrades to the Atlas V core stage are necessary, and abort from the Delta IV exceeds allowable g-loads. In the end, the probabilistic risk assessment…indicated that the shuttle-derived Ares I was almost twice as safe as that of a human-rated EELV."

Steve Cook, director of the Ares Project Office at NASA's Marshall Space Flight Center,[12] agrees. "The Atlas V and the Delta IV were designed as a low cost system to get cargo into space. It's about a wash when you compare costs to modify Ares I for a crew, but Ares I is much safer and more reliable because you have fewer propulsion systems—two versus four in the case of Delta IV—and the system is already designed for a crew. The EELV family just doesn't lend itself well to growing into a system that can throw 300,000 pounds into low Earth orbit. In a sense, we've pulled in the best from Delta (the RS-68 engines), but transforming the system of EELV's into Earth departure stage is not practical. It doesn't get you where you need to go, ultimately. "

Costs turned out to be the final nail in the EELV coffin. Studies showed that the cost of an EELV-based launch system was nearly 25% higher than the Ares I and V boosters. Griffin concluded, "While we might wish that 'off the shelf' EELVs could be easily and cheaply modified to meet NASA's human spaceflight requirements…the data say otherwise."

Once EELVs were out of the running, it was time to consider other approaches. Steve Cook tasked his team with evaluating space shuttle main engines. The SSMEs are made more complex by virtue of the fact that they are designed for reuse. On Ares, these engines would be used only once. In short, Ares main engines did not require the complexity of the reusable SSMEs. So designers turned to the tried-and-true J-2, the engine that powered upper stages of the Saturn V through a decade of successful flights.

12. Marshall Space Flight Center is in Huntsville, Alabama.

The Delta IV EELV, built by Boeing. (Photo taken by Carleton Bailie and courtesy of United Launch Alliance.)

Lockheed Martin's Atlas V EELV. (Photo courtesy of Lockheed Martin.)

"Our baseline [study for Ares I] called for an upper stage with an SSME and a first stage that used a four-segment solid rocket booster similar to what we use today on the shuttle," Cook explained. "Ares V had a five-segment solid booster, and a core stage with five SSMEs." The baseline second stage, which was the Earth departure stage, would have a single J-2 derivative. J-2 won on the merits of cost, reliability, and safety. But as Cook's team moved beyond the initial study, it became evident that Ares I and Ares V had far too many different propulsion systems, including solid rockets, strap-on solids, the Ares V's SSMEs, and Ares I's J-2 engines. Cook wanted to minimize the number of developments required for new propulsion technologies. One way to streamline the system was to get more commonality between the Ares I and Ares V.

Designers settled on the J2-X, an advanced version of the *Apollo*'s J-2, for both the Ares I upper stage and the Ares V upper stage. The J-2 was less powerful than the SSME, so planners needed more power from the first stage. They got it by expanding the stage from four fuel segments to five. This made the size of Ares I's first stage identical to that of the Ares V, so both launchers shared common hardware. This not only saves money in manufacturing but in processing as well. Building and launch facilities now could share common size for the first stages of both Ares I and V. But, according to Cook, there was still an issue. "We still had SSMEs running around on the first stage of the Ares V, so we said 'how can we get rid of those?' Some of our guys got really creative and said, 'We'd really like to use the largest liquid oxygen/hydrogen engines commercially available today, the RS-68 (used on Delta IV heavy lift vehicles). It's a lot cheaper than the shuttle engines; it's proven.' The problem was that those engines weren't giving us enough power." The solution was to give the engines more propellant. Ares designers scaled up Ares V to be 33 feet in diameter (rather than the original 27.5 feet), allowing them to put more propellant on board. This gave Ares V a diameter nearly identical to the Saturn V, and the new engines will actually perform better than the original projections for SSMEs. Just as important to the budget, launch processing sites such as Kennedy and Michoud [where shuttle external tanks are processed] still had structures originally scaled for Saturn V *Apollo* Moon rockets, so the new Ares V would fit without extensive modifications. "Now we have direct traceability from Ares I to Ares V in two key propulsion systems," Cook explains, "and we're using a core stage for the Ares V, which is already flying today, so we won't have to do a lot of development work. In doing so, we ended up saving several billion dollars over the life-cycle of this program without compromising the safety or reliability of these systems."

It was a fast-paced, dynamic decision process, but Cook "had already looked at hundreds of different options, so we were already running at a fast pace and we just kept going." The speed with which the Ares design decisions were made reflects the pace at which the *Orion* project—and the Constellation

Designers of the Ares family have selected upgraded J-2 engines, whose heritage reaches back to the upper stages of Saturn V, like this one on the Saturn's third stage. (Photo by author.)

program in general— is progressing. Johnson Space Center's Wendell Mendell explains that NASA Administrator Mike Griffin "felt he needed to get things embedded and going during his tenure, so Constellation was born." The pace was fast and steady, akin to the *Apollo* days, Mendell says. "Jeff Hanley was put in place as the head of it and designed the organization after the *Apollo* management organization, which was very successful."

Having reliable human access to space—with the flexibility to use the transportation system either for people or for cargo—is a complex and difficult goal. But it is essential that Ares affords reliable, consistent, sustainable access to space, particularly far-off destinations such as the Moon and Mars.

Ares I and V compared to the shuttle and Saturn V. (Photo courtesy of NASA.)

Critical elements of the Ares I, which will carry the Orion CEV to orbit. (Photo courtesy of NASA.)

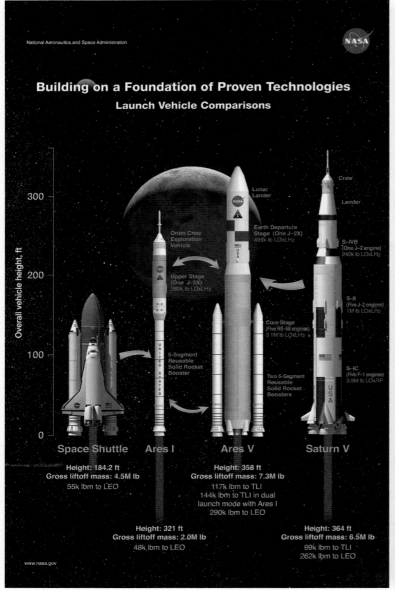

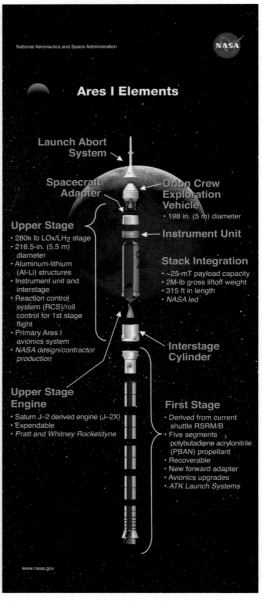

"The Moon is our first deep space frontier," says NASA's Jim Garvin. "We are really trying to build something very new, and it's going to be revolutionary. We're putting in place the capacity for humans to have access anywhere on the Moon. Anywhere."

Ares goes a step beyond the *Apollo* era in many ways, not the least of which is that it must be done with a significantly thinner slice of the budget pie than the shuttle was. Constellation is a generation beyond *Apollo*, Jim Garvin says, "particularly because we have to do it for less money. This is a different environment, a different climate. We have climate change on our planet; there's also climate change in space policy. That environment dictates that NASA has to be smarter, and better and more creative."

Part of that creativity plays out in the way elements of Constellation are developed and tested, according to Marcia Ivins, head of the Exploration Branch of NASA's astronaut office, and a veteran astronaut of six flights. "When we build this mass simulator that will be the second stage of [the first Ares test flight], they're going to assemble it in the Vehicle Assembly Building, and what they don't want to do is have to reconstruct all the platforms around it. They've figured out a way to assemble

ATK technicians prepare the new Ares I rocket first stage segment for launch. (Photo courtesy of NASA/MSFC.)

it from the inside out. You build a ring, crawl inside it, and bolt it all together from the inside out. There's a ladder structure on the inside, so you climb up and out and build the next level."

To keep costs down and efficiency up, NASA has adopted a test-as-you-go approach for much of the project, Jeff Hanley says. "We start with the end in mind. Ultimately the end that we have in mind is a human Mars mission. So we work backwards. What do we need to learn at the Moon? True, we can get back in three days, but we can put our systems out there and actually run them for a long time in an alien environment and see how they really perform. All of the tools and techniques that we develop, are those really, in practice, effective, or is there something about it that we don't understand? Then, we apply the developing technologies along the way."

This approach has led to the framework for the first full-up launch test of the Ares I, the Ares I-Y, the maiden flight of which is scheduled for early in 2013. Ares I-Y will focus on the first stage's flight characteristics, controls, and the critical separation of the first stage from the second one, which, in future flights, would carry the crew. Flight engineers will use a fully functional first stage. The stage is based on the shuttle's solid rocket boosters, which are strapped to the sides of the large external tank. In the case of Ares,

The Grand Plan: Second Generation

The *Apollo* Project used lunar rendezvous to get crews to the Moon, launching all vehicles on the same Saturn V booster. The new Constellation strategy bears closer resemblance to the original Soviet plan, in that the crews will use Earth orbit rendezvous to get the job done. For the new generation of lunar explorers, the Ares V booster will carry the *Altair* lunar lander into Earth orbit. *Orion* links up with the upper stage and *Altair* after a separate launch aboard the smaller Ares I. Once *Orion* and *Altair* are safely docked together, the upper stage of the Ares V sends them toward the Moon. In *Apollo,* the main engine of the *Apollo* CSM did all the work, settling the CSM and LM into lunar orbit. It is the powerful lower stage of *Altair*—the lunar lander—that will slow the *Orion/Altair* stack into lunar orbit. *Orion* carries far less fuel and will use its engine only to get crews back home from lunar orbit after *Altair*'s ascent stage returns crews from the surface.

the booster will be a "single stick," a configuration never flown before. The second stage is called a "mass simulator," but instead of dead weight, the stage will carry sensors to detail the flight path of the Ares I stack. As Steve Cook puts it, "There is nothing better than getting flight data as early as possible. For the first time, we will do a development test flight of our launch vehicle early enough in the development of Ares I to inform the design. That's what the Ares I-Y is all about. You can't beat flying in the environment that you're going to operate in, so we said, 'What can we do early? We're not going to have an upper stage available, we know we won't be ready with a full five-segment booster, but we can still reduce a lot of our risk early on in the project, as they did with Saturn. We looked at why they ran these tests and why they flew these vehicles the way they flew them, and it made a lot of sense to us so we took that approach. We've taken a page from the Saturn V playbook."

Another cost-saving feature of Ares I is reusability; the first stage uses the shuttle's solid rocket booster technology, developed and built by ATK launch systems. As such, the lower sections of Ares I are recycled for future launches. The Ares I first stage separates from the upper stage at an altitude of 189,000 feet, roughly 126 seconds after liftoff. After freefalling to 16,000 feet, it deploys a small drogue parachute, halving the stage's 400 mph rate of descent and tipping the booster into a vertical position. At that point, three main parachutes, each 150 feet in diameter, open to carry the booster safely to waiting recovery ships below.

Boeing is heading up the Ares I's upper stage, which can carry a 25-ton payload into Earth orbit. Boeing is also building instrumentation for the booster based on its extensive experience with the Delta IV launchers. Transition from Delta IV construction to that of Ares I will also save costs.

The Ares I is designed to eventually carry the *Orion* crew exploration vehicle, America's newest human-rated spacecraft, into orbit. *Orion* will service the ISS during the last four years of the station's lifetime, and will carry large payloads into orbit. *Orion* will also ferry crews to and from lunar orbit, but to carry out that role will require Ares I's massive sibling, the Ares V. The maiden flight for Ares V is now projected for 2018.

ORION: THE NEXT SHUTTLE

The Rocky Mountains serve as birthplace for a high-tech progeny. It is here, nestled in the unlikely setting of deer-trod foothills and granite peaks, that Lockheed Martin Astronautics has built launch vehicles, Mars orbiters and landers, and a host of defense systems. Now, LMA is producing the *Orion* crew exploration vehicle (CEV), replacement for the American space shuttle.

To many observers, *Orion*'s arrival is none too soon. The aging space shuttle fleet is the U. S. lifeline to space. It is the only American transportation system able to carry humans to the International Space Station. Its flexibility has enabled it to serve as an orbiting research laboratory, an interplanetary space delivery system, and a satellite rescue and repair platform. The fleet has been flying for a quarter century, and it's showing its age. Two of five orbiters have been lost to catastrophic failure, *Challenger* during launch and *Columbia* during reentry. The loss of human life was emotionally devastating, not only to NASA but also to all of the United States. Technicians must constantly scour the remaining three orbiters, looking for stress fractures, metal fatigue, and other safety hazards that naturally occur in an elderly flight system.

As NASA's Jim Garvin observes, "The shuttle has been a miracle of engineering, and it's done tremendous stuff, but because of the requirements levied on it after the heyday of *Apollo* it was expected to do too much. It's kind of like the Spruce Goose. It had too many things to do rather than a focus. When one tries to do that, the machine becomes very complicated."

"With the shuttle, they solved technical problems right and left in its design and development," Johnson Space Center's Wendell Mendell observes, "but they neglected to worry about what the operations costs of the final vehicle would be. If you look at drawings of the shuttle from the seventies, you see shuttles that look like airplanes with about six people walking around them, but if you look at the picture today in the bay, you can't actually see the shuttle because it's covered by scaffolding with armies of people around it like ants, doing things. That is one of the reasons it is so incredibly expensive to operate. They're working hard to make [the vehicles of Constellation] not that way."

Around 2010, after 28 years of flight, the shuttle will fly its last mission. The *Orion* may fly to the ISS as early as 2015, with Russian *Soyuz* craft—and possibly European or private vehicles—filling the gap in the interim. NASA hopes to fly the return mission to the Moon by 2020.

TICKET TO THE MOON

A flight to the lunar neighborhood aboard *Orion* will be carried out in several steps. The Ares I will put *Orion* into a ballistic trajectory, so the spacecraft must use its main engine to do one burn. This circularizes the orbit. At that time, the solar-powered *Orion* will deploy its solar panels. Says Lockheed Martin's Chief Engineer and Technical Director for *Orion*, Bill Johns, "We're not in any big hurry; we have about ten minutes before we do the burn. The current baseline is that I make sure both of my [solar] arrays deploy before I do that burn. If neither one deploys, I do an abort-once-around," returning the craft to Earth.

Once the spacecraft is in orbit and determined to be healthy, it will dock with the Earth departure stage (EDS) and *Altair* Moon lander, carried into orbit atop the Ares V, which is launched separately. *Orion* secures itself to the *Altair* on the front of the EDS, and the Ares V upper stage sends the entire stack of vehicles toward the Moon. But *Orion*'s delicate solar panels must be protected from the forces of that launch from Earth orbit to the Moon, Johns explains. "You can pivot the solar panels so they can find the Sun when the spacecraft is turning. But we typically will lock them in place when we are moving. We move them at the 'shoulder' to point aft by about 60 degrees during that Earth departure stage firing. We rotate them during that high acceleration burn to limit stresses on them."

After casting off the empty EDS, *Orion* and *Altair* coast to the vicinity of the Moon. It is *Altair* that drops the two craft into lunar orbit with its huge descent stage, but *Orion* must serve as the orbital base for lunar operations. The craft will be tasked with one of two missions. The first, called a sortie mission, sends *Orion*'s four-member crew to the surface in *Altair* to carry out up to ten days of surface exploration. For this type of mission, *Orion* spends up to 21 days in orbit autonomously. The second mission class is called the outpost mission. In this scenario, *Altair* lands at the lunar base. The crew stays on the surface for six months while *Orion* flies solo in orbit, monitoring its systems and caring for itself, awaiting the crew's return and the trip home. Although the missions differ significantly in length, Johns says, "We're trying to develop one configuration to cover both the 21-day [sortie] and outpost (210-day) missions." *Orion* returns the crew to Earth, skipping on the upper atmosphere to bleed off speed before coming into the denser atmosphere. It lands using three parachutes and a series of airbags.

ADVANCES

To the untrained eye, the *Orion* looks like a step backward from space shuttle technology, an oversized *Apollo* capsule. But a closer look reveals important advances over both the *Apollo* Moonships and the shuttles.

The most obvious difference is *Orion*'s solar panels, a departure from all previous U. S. human-rated spacecraft. Like *Apollo* before it, *Orion* consists of a crew module, a launch abort system, and a service module. The robust service module houses the main engine for *Orion*'s on-orbit maneuvering, along with a different type of power regime. *Apollo* and the shuttle are powered by cryogenics—liquefied gas—that cannot be stored for long periods. *Orion* carries solar panels that will enable the craft to endure its long flight times. Mark Kirasich, Deputy Manager for the *Orion* Project, says the decision was necessary because of weight constraints and long mission duration. "Fuel cells take consumables. They're heavy. We don't have enough throw-away to toss six months of hydrogen and oxygen toward the Moon. Instead we have a reusable energy source in solar arrays." The fan-like arrays are larger, more efficient cousins of the panels used on the Mars *Phoenix* lander, so they have a good track record of development and operation.

Orion is significantly larger than earlier Moon ships, spanning 16.5 feet across (*Apollo* was 12' 10" in diameter), with 691 cubic feet of interior space. Although the cabin is roomier than *Apollo* command modules, former shuttle astronauts may find themselves feeling a bit cramped, says Lead Cockpit Engineer Jeff Fox. "You're going from an over-sized Suburban SUV to a small mini-van." But *Orion* has more tricks up its technological sleeve. *Apollo* was fitted for two-week flights; *Orion*'s solar power enables it to stay in Earth's or the Moon's orbit for six months. It can carry a crew of six to the ISS, along with supplies, or a crew of four to the Moon. The craft can be tethered to the ISS and left to fend for itself for months or can hibernate in orbit around the Moon while crews spend half a year at the lunar outpost or on long-duration exploration sorties.

These features were built into the initial requirements of the spacecraft, Lockheed Martin's Bill Johns explains. "One fundamental requirement was for it to be a lifeboat for the ISS. But we also will design the craft with commonality between lunar missions and low Earth orbit. You do six month rotations on ISS, so you say, 'We'd sure like this thing to be able to look after itself for six months, docked to the ISS. The amount of power we can gain docked to ISS isn't a whole lot different from what we can get at the Moon, so why not do six-month rotations at a lunar base?' There's a lot to be said about one design for multiple missions. Six months at ISS lends itself to a

The solar panels of the Phoenix Mars lander, seen here in July of 2008, are smaller cousins of those on Orion. (Photo courtesy of NASA/JPL-Caltech/University Arizona/Texas A&M University.)

spacecraft that can also handle autonomous flight around the Moon for six months."

Orion's automation is critical to its missions near Earth and the Moon. The Russian *Soyuz* spacecraft have had the capability to dock without human input for decades, but this is the first U. S. craft to have the capability, and its talent for untended orbital moves will be unparalleled. JSC's Mark Kirasich says, "Technology makes a huge difference. Computers and data networks are affecting things across the vehicle. Now, we have 100 megabit and gigabit speed data buses. Back in *Apollo*, it was analog. In the shuttle era, it was much slower rates with raw numbers and data values. Here, it's going to be images and plots and video. We can pump a lot of data around." As an example, Kirasich cites the shuttle's flight data file, a series of thick books like instruction manuals, a sort of user's guide to the spacecraft operations. Those thousands of pages will now be on a screen, and the same screen will enable the astronaut to execute the operations that the manual calls for.

Artist rendering of the Orion *spacecraft with deployed solar panels. (Photo courtesy of Lockheed Martin Astronautics.)*

Kirasich's team was watching closely when, in April of 2008, the European Space Agency's fully automated transport vehicle *Jules Verne* delivered cargo and fuel to the ISS. If the shuttle is vintage 70's technology, *Jules Verne* is 2000's technology, Kirasich says. "The shuttle used radar (invented in WWII), as well as optical sensors. That was state of the art in 1970. *Jules Verne* uses GPS and laser. The accuracy is phenomenal. In its final approach, the crew watches as it comes in. They have a camera watching a circle of 5 degrees. In training, simulators wander all over the circle area, which is what everyone expected. In reality, it never came off the centerline." This type of precision automated docking may be needed for some ISS missions, and will be critical for future lunar missions.

Artist rendering of the interior of Orion, *showing the four-astronaut configuration. (Photo courtesy of Lockheed Martin Astronautics.)*

In low Earth orbit, spacecraft can make use of global positioning satellites (GPS) for accurate navigation. But once outside of those orbiting satellites, the system no longer works. *Orion* crews must use another approach to navigation as they come and go from the Moon. Engineers have selected a single technology that will work in both arenas, Mark Kirasich explains. "We use our S-band system that we use for communications, but we embed navigation information into the radio signal so that the two vehicles

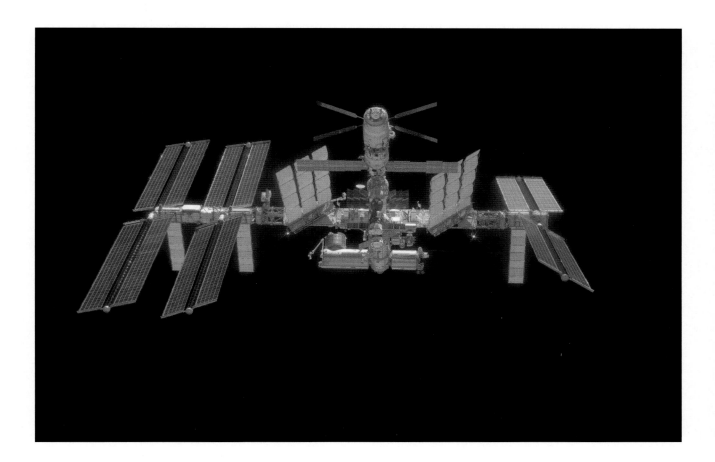

can do ranging and Doppler measurements. It's clearly a step up from what we could do in the seventies."

With its full automation, *Orion* can serve as an uninhabited cargo vessel for the ISS. It will also be capable of taking over the docking events with *Altair* landers on their return trip from the Moon, should *Altair*'s rendezvous systems fail or the crew members become incapacitated.

Bill Johns believes that temperature is as great a challenge as long-duration autonomy. He points out that commercial spacecraft have been taking care of themselves for fifteen years or more. But he says there are some unique things associated with the heating environment around the Moon that make two weeks as challenging as six months. "If you're orbiting over the poles, you are basically in the Sun almost the whole time, and you're getting solar reflection and heating coming up from the Moon, so you have a fairly hot environment that you have to accommodate. But once you size for that case, the duration isn't a long putt."

The dual nature of *Orion*—serving both the near-Earth space and lunar environment—necessitates two subtly different versions, referred to as "Block One" and "Block Two." The Block One craft functions as a crew and cargo ferry to the ISS, and carries up to six astronauts. It runs at a slightly higher cabin pressure than Block Two, matching the ISS's nitrogen/oxygen mix of 14.7 pounds per square inch (psi), equivalent to air pressure at sea level.

The European Space Agency's Automated Transport Vehicle (ATV) Jules Verne, docked at the center of the ISS. The European ATV can carry three times the cargo of a Russian Progress tanker, roughly 7.7 tons to low Earth orbit. (Photo courtesy of NASA/JSC.)

The reason: the Moon-bound Block Two *Orion*, with a crew of four, must interface with the *Altair* Moon lander, which has lower pressure and a higher percentage of oxygen. *Altair* and the lunar version of *Orion* operate at a pressure of 10.5 psi, similar to a mountain community at 8,000 feet altitude. The lower pressure enables Altair to be a much lighter craft, as it can carry less air for its long missions.

Another difference between Earth-orbiting and lunar *Orion*s has to do with the heat shield, an ablative surface that protects the crew and craft as it burns up upon return through Earth's atmosphere. Entry speeds are considerably higher when a spacecraft comes from the Moon. Designers must add roughly 500 lbs of material to the heat shield for lunar CEVs. But the aerodynamics remain the same, so adding mass to the heat shield is not difficult.

The MMU: Going It Alone

In February of 1984, Bruce McCandless became the first human "satellite," flying untethered some 320 feet from the space shuttle Challenger. Here, he reflects on that experience, and on post-shuttle advances.

The Manned Maneuvering Unit (MMU) served as a pathfinder and a demonstration for the SAFER (Simplified Aid For Extra Vehicular Rescue, now used at the ISS). SAFER is intended for self-rescue in case you get separated from the station. SAFER has versions for both the ISS and *Orion* suits, and snaps on. It's much more compact than MMU was, but it has enough gas to get you back. If you crouch down and shove off of the station, you end up with about a 3 feet per second velocity. SAFER is designed to take that velocity out and get you back. It has about an 8 foot per second change in velocity. It's just to get you back to where you can grab something.

During my MMU test flight, I had anticipated some solitude and being able to turn my back to the shuttle to gaze out at creation. I never got the chance. The communications were too good. We had three channels going at once. I had Commander Vance Brand reminding me not to go too far away, and not to go under the wing where he couldn't see me, and to stay away from the engines. Then there was Mission Control wanting to know how much nitrogen [fuel] and battery power I had left. And [fellow astronaut] Bob Stuart wanting to know 'When's my turn?' In the middle of all that, I never really got the chance to stop and do a Walden Pond type thing. But it was very impressive and beautiful. At one or two points I got to look down. I had no idea where we were until, at one point, I looked down and saw that we were right over Florida. You cannot mistake any other place on Earth. It was reassuring to see the Cape [launch] complex, and the Florida Keys and the Bahamas.

For the next generation of [spacecraft] software, we've come a long way. To take it to a down-to-earth level, one of my nieces gave my wife an iPod for Christmas with two gigabytes of flash memory. If you clock back to the late seventies when I got my first laptop, it was a Zenith: twin floppy disk drives, a bit bulky. After a year, I finally had saved up enough money to buy one megabyte-worth of additional ram. It cost $1,000. If you scale that, this two-gigabyte iPod would have been a $2 million purchase, and you would have needed a little trailer to carry it around. We've made such fantastic strides. (*Apollo* astronaut) Charlie Duke's son is flying the 777. They make absolute zero-zero landings hands-off. My understanding is that they have a three-channel autopilot system and a three-channel ILS altitude type system, and when the weather gets crummy, you approach until you're on the ILS glideslope, and you engage the system and put the gear down. As long as you have six green lights, you sit there with your hands folded while the system takes the airplane in, lands it, and throttles back and applies the brakes and says 'Here you are.' There is no reason why *Altair* shouldn't have triple or quadruple redundancy. On *Altair*, the basic decision has been to wait about ten years and see what happens to electronics in the meantime. That's probably a pretty good plan. One of the things that we continually do to ourselves is we lock something in, and by the time we go fly it's antiquated. Last fall, there was some complaint that somebody had hacked into the e-mail system on the space station. NASA put out a call for help to Microsoft, and Microsoft replied, 'Gee, we'd really like to help you, but all the people that are familiar with Windows 3.1 are retired, and believe it or not, our people are really gainfully employed debugging Windows Vista right now.'

AT THE CONTROLS

Orion's flight deck is like nothing before it in human spaceflight. Instead of bulky switches and dials, the primary interface for pilots consists of three flat screens, similar to cockpits in today's 787 commercial airlines. The arrangement and layout of the modernized controls is the responsibility of Johnson Space Center's Jeff Fox. "We're looking at taking the twelve or fifteen hundred switches that are on the shuttle—taking all those manual control points, and putting them in the software. We've got all that contained in three pieces of glass." Each 10" x 8" horizontal screen is split into two areas, so astronauts will have a total of six screen areas in which to carry out diverse functions. "We'll have maybe fifty switches. Everything else is in the software."

Fox's cockpit working group must take into consideration *Orion*'s operations, engineering, human factors, and life sciences. "We're looking at every aspect of what the crew touches in the pressurized volume," Fox says. "Displays, windows, lights. How are things laid out, how you strap in, how you do your procedures, because it's no longer paper, its electronic. It's a huge integration job." Fox and his team of engineers have studied past spacecraft, including shuttle, *Apollo* and *Soyuz,* and have consulted with airline companies to come up with the best arrangements for *Orion* crews. "We've looked at all those spacecraft. We've been in the *Apollo 17* command module here at Space Center Houston—I've lost track of how many times—climbing around in there, thinking about what they did and how they did it, talking to the astronauts and the *Apollo* Human Factors/Habitability folks. Then we compare to what we've got on shuttle."

While spacecraft architects toil over computer screens, Fox's team builds out of plywood and plastic. "You need to have a physical environment," Fox asserts. "There's only so much you can do on paper. We have found that the [computer] modelers come over and verify how we laid it out, and go back to make changes. There is a lot of give and take."

Fox's fabricators built three different venues. The first was a roughed-in, low fidelity foam-board version to ascertain gross placement of systems and crew. A second is medium fidelity. It uses real adjustable seats and is an aluminum structure instead of foam board. Designers are able to build operational workarounds, making changes early and cheaply. A third mockup is chopped off just below the crew deck. This one is a physical study of tanks, boxes, access panels, and plumbing. The mantra of crew safety is everywhere: seats and other structures are outfitted with struts, braces, and shock absorbers to attenuate the landing jolt. To Fox, it's a game of trying to outsmart the things that can—and often do—go wrong. "You come in at an angle, and that attenuates the [landing stress]. But what if the angle isn't perfect? What if you hit the side of a wave? What if you wind up on the land and there's a burm instead of a flat space, or what if you get a damaged parachute? All kinds of things can come up, so how do we protect the crew better?" Fox points to a seat resembling something that Danica Patrick might use, with racecar-like lateral support. "We'll have something to help

keep the head from moving too much, and to support the spine in relation to the body. We teamed with the racecar industry to build conformal seating. If we do that, maybe we won't need struts on the couches [which was the approach in *Apollo*]."

Hanging from the ceiling are ping-pong balls on strings. Capsule builders use them to get the right "eye point" for test subjects in the couches. It's important, Fox says. "If you're not in the right sweet-spot, the windows don't work, the displays are all in the wrong place; the reach, the visibility, the access to critical areas all has to work. When you talk about something like changing the seat thing, we always say, 'be mindful of the eye-point.'"

The low-fidelity mockup of Orion has rudimentary couches and is useful for fleshing out the interior space of the CEV. (Photo by the author.)

Mockups enable crews to test visibility and placement of windows. Note the ping pong balls attached to the upper sill of the window frames. These help designers locate the best position for crew eye placement. (Photo by the author.)

In the crew exploration vehicle design business, everything is a trade. If CEV builders push one thing, something else pops out and it changes the overall shape. Power, weight, functionality, habitable volume: every time designers tug on one, it affects something else. The system must also be

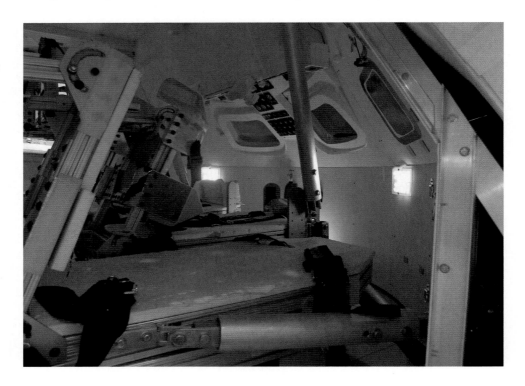

Orion mockup used for arranging systems carried below the flight deck, adjacent to the heat shield. (Photo courtesy of Marianne Dyson.)

flexible in all sorts of flight regimes. The crew may be restrained in a suit, operating systems during launch. They may be experiencing extreme vibration of spacecraft maneuvers or in crisis situations, in pressurized and depressurized conditions. Fox must envision all the possibilities. "It's got to work in an emergency. Say I'm in an emergency entry, so my suit is puffed

Technicians use cables, hoses, and boxes to size the interior cabin space of Orion. Note manikins for couches and inflatable figures for overhead surfaces used in weightless conditions. (Photo by the author.)

up now, and I'm all strapped in real tight because I'm trying to protect myself, and I'm not really going to be able to reach up [to the screen panel] to do this kind of stuff, so I have to have a device down here by my gloved hand—like a cursor or a track ball—that can interface with the software that's on the display." Screens will probably not be touch-screens, because in the microgravity of space they can be bumped. Instead engineers are considering bezel keys, trapezoid-shaped buttons similar to those used on commercial airliners.

Although *Orion* may seem cramped by shuttle standards, its quarters will never be inhabited for more than sixteen days. In lunar orbit, when the vehicle is in an automated configuration, the ground will have a great deal of control. They'll be able to look at all the data in the vehicle, talk to it, and monitor certain automated systems. The vehicle is out of contact for half of every orbit (when it passes behind the Moon), so it has to be smart enough to take care of itself. Astronauts on the lunar surface will make sure it's a "good vehicle" before returning to it.

REUSABILITY

Unlike *Apollo*, elements of *Orion*'s crew module may be used up to ten times. But the reuse of space flight-worthy equipment requires a fairly delicate landing. Originally, the intent of CEV designers was to end each flight on dry land, with water landings only as an emergency contingency. Studies indicated that a dry landing would enable the reuse of the entire outer structure of the CEV, along with about 75% of the overall spacecraft components. But a lot can go wrong when you land on dry ground. Mark Kirasich's engineers considered the stresses on the spacecraft if one of the three parachutes failed. What damage, they wondered, would the spacecraft sustain during high winds, or landing on steep slopes? "We found that we needed a very robust system so that we could still end up reusing the spacecraft."

The initial solution engineers came up with was to deploy airbags with a parachute system. To preserve the spacecraft, designers considered a rugged airbag system around the entire vehicle. The bags would have deployed in a cushioning ring around the entire heat shield, which would need to detach in order to free the bags for inflation. "That adds complexity," Kirasich says. "You've got to have a mechanism to blow the entire heat shield off—and a lot of mass. We just couldn't take all that mass to the Moon and back."

The Russian space program has landed its crews on land for decades. But the difference is that *Soyuz* spacecraft don't go to the Moon, and they are not reused. The design team backed off and went for a water landing, but even that requires at least some minimal design to cover contingencies involving land landings.

Water, especially salt water, changes the reuse equation dramatically. Although the goal for *Orion* was 75% reuse, ocean landings pushed the figure down to 20 or 30 percent. Bill Johns says, "In a water landing, it's very difficult to be able to seal everything against the saltwater and air. They start to do their work electrochemically on the surface. That can ultimately result in microcracks. You just can't get to every exposed surface to clean it out. For the nominal water landing, we can protect everything inside of the pressure vessel, but the pressure vessel itself is difficult to preserve." Adding to the problem is that after several hours in the water, *Orion* crews would begin bringing fresh air in with a snorkel, infusing electronics and materials with damaging salt fog. After being soaked in seawater, the craft would then sit on the recovery ship for several more days, in the salt air. "That gets expensive. We're throwing away half the cost of the spacecraft each time," says Kirasich. Waterproofing the spacecraft by putting coatings on the metal, sealing certain interfaces, and keeping the hatch closed as long as possible brought reusability up toward the 50 percent mark. But the cost of a water landing still seemed too high.

Engineers went back to the drawing board and came up with a design solution for landing on land. After looking at crushable structures, different types of seats and other parachute designs, Bill Johns' team came up with a modified "toe" airbag system—so-called because it wraps around the leading edge—that would not require detaching the heat shield. Johns describes the process: "We scratched our heads for a couple weeks and said, 'What is it, short of a full airbag system, that would make sense if you're going to drift over land?' That's when we conceived of the toe airbag system." By adjusting the parachute risers, the CEV hangs at an angle of 28°. Airbags deploy only on one side of the craft, exiting through a panel in the side. Airbags can inflate in stages, so that they wrap around the edge of the spacecraft like a chain of grapes. These are bags within bags, so that the outer bag vents upon impact to avoid bounces. The additional benefit is one of weight savings: the toe airbag system is less than half the mass of the earlier study.

The elements of water, air, and earth are not the only dangers facing the delicate workings of *Orion*. Another is vacuum. While salt water corrodes materials, a vacuum tends to preserve materials. But *Orion* must be prepared for any emergency, including the loss of air in the cabin. A failure in a pressurized tank, an explosion, or a micrometeoroid hit could contribute to a deadly loss of pressure. Researchers projected a certain size of hole that is most likely to result from such a failure. Engineers were then tasked with designing the pressure vessel of the spacecraft—the portion housing the living and working areas for the crew—in such a way as to hold the pressure for 45 minutes to one hour. Jeff Fox outlines the scenario: "What if I'm up in orbit and all my stuff is stowed. My seats are behind these panels. I have access to the storage under the floor like food and laptops. So I've got everything out, and then I get a leak in the cabin. In a certain amount of time I've got to get back into my suit and get the seat in. So I get in my suit and put my umbilical on, so now I've got all these umbilicals all over. The suit's starting to puff up a little because the pressure's dropping. Now I've got to maneuver around everybody and put my seat back together."

Once the spacecraft pressure drops down to the vacuum of space, there is a new challenge. All the electronic boxes that relied on air circulation to keep cool are now in danger of failure from their own heat. "Those flat panel displays are going to overheat just like that," says Johns. "So everything inside that pressure vessel that has more than about 15 watts going to it is all going to be on cold plates. We've had to design everything [that dissipates heat] to be on a cold-plate so you have a way to remove the heat in a vacuum." As *Orion* moves from its preliminary design phase to the critical design phase, in which 90 percent of the actual blueprints are completed, engineers are hopeful that they can hit the 75% reusability mark given to them at the start.

THE SCHEDULE, FOR NOW

The first production CEV will launch on Ares I-Y, the first full-up test of the Ares I booster. It is unmanned and will demonstrate a high-altitude abort and water landing. This *Orion* will not have a service module. It was scheduled to launch in September of 2013 and has already suffered several delays due to budget and scheduling issues. The next flight is also uncrewed and will conduct most of an ISS mission. Planners are still debating whether it will dock or not. The third mission carries two crew members and will dock to ISS. It will deliver the next-generation docking adapter called the low-impact docking system, or LIDS.

LIDS was developed for the entire lunar architecture. Bill Johns describes LIDS as "a common, very efficient, and very mechanically redundant docking system." The ISS docking system currently in use is the Russian APAS (the androgynous peripheral adapter system). The first time *Orion* shows up at the ISS, it will carry an adapter with the APAS docking system on the front side to attach to the ISS port, but when it separates, it will leave behind the new LIDS system on the exterior. The first two missions will leave this new connector on two docking ports now reserved for U. S. craft. At this point, ISS will be able to completely interface with NASA's next generation of Constellation space vehicles.

Concurrent to, or shortly after, the first two *Orion*-ISS missions, the *Altair* Moon lander will carry out uncrewed tests. Current plans call for *Orion* and *Altair* to carry out the seventh lunar landing sometime in 2020. In the forty-some years that have passed since the first landing, materials, strategies, and technology have changed. Lockheed Martin's Bill Johns is amazed by what has come before. "What I learn, every day, about what we did forty years ago is really impressive. Their big challenge was that they were the first to do it. Our big challenge is to do it in a sustainable way. It's all about sustainable human exploration."

THE ALTAIR MOON SHIP

If aerospace is a game of Monopoly™, Lauri Hansen has a get-out-of-jail-free card. With the work ahead, she may need it. Hansen is the Lunar Lander Project Manager for NASA. Her assignment: design a vehicle that can be

European Partners

The European Space Agency (ESA), a community of 17 spacefaring nations, is engaged in an ongoing study of projects and advanced technologies that could support a human-inhabited Moon outpost. A recently completed assessment called the NASA/ESA Comparative Architecture Assessment resulted in detailing concrete ways in which NASA and Europe could collaborate on various scientific and exploration scenarios. ESA has built a rich heritage of human spaceflight experience with the shuttle and ISS, including its massive *Columbus* station module. ESA has also developed the cargo-carrying *Jules Verne* automated transport vehicle (ATV), which has direct applications to future automated cargo capabilities. Europe's largest aerospace company, EADS Astrium, recently unveiled another variation: a *Jules Verne*-style ATV that could carry 3-person crews on lunar missions.

ESA is considering, in detail, such concepts as a lunar cargo landing system to be launched aboard the *Ariane V* (ESA's largest commercial booster), European communication and navigation systems for spacecraft, and lunar outpost elements, ESA-developed human-rated craft that would launch aboard *Ariane V*, orbital outposts, and lunar surface habitats and rovers. In a recent ESA release, ESA Exploration Program Manager Bruno Gardini said, "After the satisfaction of the successful deployment of *Columbus* and ATV we are looking forward to enhancing our role in the partnership for a sustained and robust space exploration program where human spaceflight is the cornerstone. The Moon is surely an important case study and a useful testbed to thoroughly prepare for more distant destinations."[13]

Michael Bosch, president of the Hamburg University of Applied Sciences, is part of a fourteen nation International Space Exploration Coordination Group, which includes member nations of ESA. Bosch says, "NASA's strategic transportation infrastructure does not allow for international work, so ESA and the Russian Space Agency are studying options based on their own launchers. The goals of the ISECG include sustained and self-sufficient human presence beyond Earth orbit. We are after interoperability between systems." Strategists are considering upgrading ESA's ATV to carry a crew of three. This would involve adding a return module with heat shield and an escape system for the launch phase. First launch could be as early as 2013. "ESA believes Europe should have its own human exploration infrastructure with full access to the Moon and Mars." This access would provide redundancy of human access, a backup in case of failure or delays in other projects, and the capability of international rescue operations. A second option under study is called CSTS and would modify a *Soyuz* for a crew of six. It would launch on a Soviet booster from Baikonur. Recent meetings have resulted in the tabling of Europe's involvement in this option, at least for the foreseeable future.

13. Excerpts of this July 9, 2008, release can be accessed through the ESA exploration portal at www.esa.int/exploration.

launched atop a booster that does not yet exist, make that vehicle compatible with another spacecraft that is still in design stage, and build enough flexibility into the lander's nature that it can transport tons of cargo to the lunar surface to build a lunar outpost whose plans are amorphous at best. Astronaut Marsha Ivins comments, "Laurie's project gets a pass on some of the constraints so she can think outside the [corporate] box, whereas the more classically run projects [like *Orion*] are constrained by the box."

Hansen's project has been named Altair. It's a moniker full of symbolism. "Altair is the eleventh brightest star in the sky," she says. "The star's name comes from an Arabic phrase for 'the flying one.' Altair is in the constellation of Aquilla (the Eagle), so it has a nice futuristic feel but acknowledges our heritage back to *Apollo*."

The symbolism is fitting: Altair has some large and complex shoes to fill, serving several roles in the Constellation architecture. To fulfill its mission, the craft must be massive. While *Apollo*'s Lunar Module was designed for flight to and from the lunar surface from lunar orbit, Altair's huge descent stage must slow the *Orion*/Altair stack into lunar orbit as well as taking payloads to the surface. This added role requires a great deal of fuel, especially when the landed cargo capacity of the craft approaches 17 metric tons. While *Apollo*'s Lunar Module stood at a height of 21.3 feet (7 m), Altair will tower over the lunar surface at a height of 32.5 feet (9.9 m). Most of the height is in the descent stage. The top of the descent stage, where astronauts will step out from their cabin, stands 6 meters above ground. The deck is just under 9 meters across.

Altair must serve three roles, each unique.

Sortie Variant

The first form Altair takes will be the sortie variant. While *Apollo* ferried a crew of two to the Moon for stays up to three days, and had a range limited to equatorial regions, Altair's sortie variant provides a crew of four with access to the entire globe for missions lasting up to seven days. These sorties will provide scientists with access to rugged highland areas and polar regions never before accessible. Constellation's Jeff Hanley says, "With a crew, it can take a 2 ton payload. *Apollo*, at maximum, took 500 kg, so we're talking about up to 4 times that, to places *Apollo* couldn't get to with twice as many crew for twice as long." Designers liken the sortie variant to a camping trip, where crews live out of the module for a short period.

Apollo 16's Charlie Duke spent three days on the lunar surface. For a longer stay, he suggests several areas important to designers. "A good operational layout of the crew module is important: ease of operation, ease of systems operation and maintenance, handling of emergencies. Good visibility is another consideration. Then once you land four folks in that thing, you've got to think about habitability. How do you sleep? Where's the stowage? You think about ingress and egress through the airlock. Those practical design elements make things livable and doable."

When Altair launches, it leaves behind the descent stage as well as the airlock. Elements of both may be reused, depending on their design and robustness.

OUTPOST VARIANT

Altair's second identity is that of an outpost variant. Its role is primarily transportation, a Moon bus to get the crew from orbit to a settlement. Unlike the sortie variant, this craft does not need an airlock, as the crew cabin is small enough to decompress easily. Instead of living in the Altair lander, the outpost variant sees a crew exiting once to live at the outpost. The small crew cabin and less crew-supporting supplies frees up space for more cargo. The Altair would hibernate on the surface for up to six months before taking a crew back up to an unmanned *Orion* for the return home. "In some ways, this is the most difficult from a design standpoint," Hansen feels. "It has to sit dormant for six months." Keeping a complex, untended craft healthy in the lunar environment for a long period will require more insulation and more power, and systems will need to self-evaluate periodically. Some lander

protection might come in the form of deployable tents, stored at the base and reused for each new outpost Altair vehicle.

Cargo Variant

The third Altair type is the cargo variant. As its name implies, this craft's sole purpose is a one-way supply trip to the outpost. Instead of a crew, the fully automated ship would transport up to 17 tons of cargo to the surface. Typical flights could carry entire habitats, rovers, construction vehicles, oxygen, and other consumables, or heavy equipment to support power or communications.

The descent stage on all three Altair types is powered by cryogenic (super-cooled liquid gas) fuel. The prime candidates for fuel are a combination of liquid hydrogen and liquid oxygen. Cryogenic fuel has more power than types of fuel that can be stored at lunar temperatures. This bigger bang for the buck comes with a price: fuel must be refrigerated, and that costs power. The ascent stage cannot afford to have cryogenic fuel, as it will stay on the lunar surface for weeks or months before being ignited to return. Refrigeration over those long periods is not practical. Instead, proven hypergolic[14] fuels—fuels that

Orion *and the* Altair *lander in orbit around the Moon. (Photo courtesy of NASA/Lockheed Martin.)*

14. The term "hypergolic" refers to fuel that ignites upon contact with an oxidizer. Hydrazine and nitrogen tetroxide are commonly used in spacecraft.

Moscow's Moon Tourism

Perhaps in response to America's new lunar plans, Russian companies are again exploring the possibility of space tourism, this time to the Moon. The goal would not be to land but rather to circumnavigate the Moon. The Moon tour is envisioned by designers at RKK Energia, the company that builds a family of successful launchers and the *Soyuz* and *Progress* spacecraft. A retooled *Soyuz* spacecraft would carry a three-person crew—one of which would be a paying tourist—coasting in a free-return trajectory around the far side and back again. The European Space Agency is studying involvement in the project, including the manufacture of a habitation module based on its designs of the *Columbus* space station module and the *Jules Verne* resupply craft.

The trip is envisioned as taking place after the *Soyuz* completes a routine service call to the International Space Station. Some elements of the Soyuz would be modernized and upgraded, like communications and heat shields. The craft would need added power to leave Earth orbit, so it would meet up with a booster stage, launched separately.

Once in orbit, the two craft would link up for the lunar journey. Reservations will be required in advance.

Beyond tourism, Russia is studying the ACTS spacecraft, a *Soyuz*-derived lunar lander that could carry a crew of six to the lunar surface. Joint discussions between potential Russian and European partners are currently on hold.[15]

---

15. See the Planetary Society's *Planetary News: Spaceflight, Europe and Russia Join Forces to Study Advanced Crew Transportation System* by A.J.S. Rayl, June 28. 2008. http://www.planetary.org/news/2006/0628_Europe_and_Russia_Join_Forces_to_Study.html

The Russian Moon orbiter ACTS. (Art ©Anatoly Zak/Russian SpaceWeb.com)

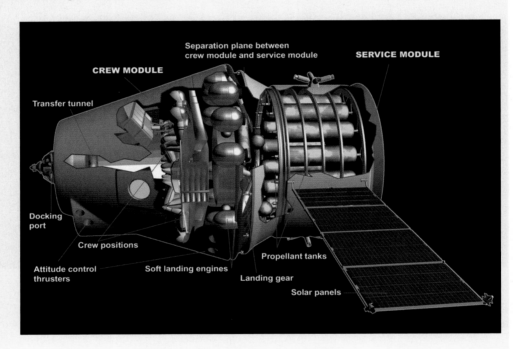

can be stored without refrigeration—will be used. Although less powerful, they also take up less room and are stable over long periods.

At this stage of development, all scenarios are being pursued, from the practical and mundane to the strange and creative. One future variation under consideration is a sortie *Altair,* whose descent stage is equipped with wheels and navigation equipment. This scenario envisions a future science expedition voyaging north of an already established outpost at the south pole. This sortie mission descends from orbit to its scientific target area. The crew carries out a week-long exploration and then returns to the *Orion* CEV in orbit overhead, leaving the descent stage as usual. But this descent stage has a different bag of technological tricks. The craft points itself south, heading toward the Shackleton base while gathering data along the way. When it arrives at Shackleton, the outpost now has another rover or cargo carrier. Marcia Ivins believes that mobility is a key to efficiently building an outpost.

Google Space

The latest—and most well-funded—private sector attempt to generate interest in the Moon comes from the Internet giant Google. A total of $30 million is up for grabs in Google's Lunar X-Prize initiative. The rules stipulate that the entrant must be at least 80% privately funded. To win, the team must successfully land a rover on the lunar surface, and that rover must travel a minimum of 500 meters. Rules also call for video and still images to be transmitted to Earth. The grand prize of $20 million covers the successful roving mission, but bonuses are to be had. If the craft images man-made artifacts such as *Apollo*, *Surveyor*, or *Luna* landing sites, the team will receive an additional $5 million bonus. An additional $5 million second place will also be awarded. In a Google Lunar X-Prize press release, CEO Peter Diamandis said, "The Google Lunar X-Prize calls on entrepreneurs, engineers, and visionaries from around the world to return us to the lunar surface and explore this environment for the benefit of all humanity…We are confident that teams from around the world will help develop new robotic and virtual presence technology, which will dramatically reduce the cost of space exploration."

Altair project manager Laurie Hansen says, "Any advances in industry—particularly small companies—helps feed the excitement, it helps feed advances with good ideas. We haven't seen anything drastically new yet, but just the fact that it's feeding the thought process and getting everybody excited is great." And while Google's competition promotes advances in technology, its cultural implications may be even more important, Hansen believes. "NASA has moved from a very small engineering organization without a lot of process controls and so on—which is the way people envision the *Apollo* days—to this big monolith of getting things done. There's always a happy medium somewhere. As you add bureaucracy, you lose some things, and you have to keep asking what can you learn from the smaller guys? Frankly, they can take a lot of risks that we can't. If they go and build a lander and it crashes, as it did recently with the X-Prize contest, everybody says, 'Well, they gave it a good shot, and man were they close.' If NASA builds a lander and it crashes, that's not the reaction that we're going to get. It used to be that way, back in the good ol' days." Hansen suggests that in the *Apollo* era, people understood that the space program was experimenting, pushing the envelope, and that the essence of this exploration was not only technology advancement but danger.

Today, the culture at NASA emphasizes risk management and astronaut safety. With the loss of two shuttles, many feel these attitudes are prudent and reasonable. The direction NASA takes is largely dictated by social and political mores. Whatever the drivers, some analysts believe NASA has lost momentum in terms of the kind of dramatic exploration that leads to great discovery. Diamandis wants to change that equation, not only by inspiring entrepreneurs, but by feeding new technology and design into the pipeline where NASA—and the rest of the world—will benefit.

"At the end of the day, I've built an outpost and I ask how many missions did it take me to do it. If I can drive the parts around, that's fewer parts, ultimately, to send, and the cost of the project goes down."

HAPPY LANDINGS

The *Altair* crews will face challenges that *Apollo* crews did not. Although the wide rim of Shackleton crater is fairly smooth and rounded, providing a large landing area, the surrounding terrain is rugged. Many shadowed craters will spread a confusing landscape below the astronauts piloting their landers. Adding to the visual confusion will be low lighting angles. The long periods of solar energy for the base also mean long shadows. *Apollo 14* Lunar Module pilot Ed Mitchell contrasts *Apollo*'s landing conditions: "We were trying to land such that the Sun angle was equivalent to seven o'clock in the morning. The Sun was at eight or ten degrees [above the horizon]. The fact that you have a long shadow is very helpful in the landing process. We used the long lander shadow to help with depth perception, as well as using the altimeter."

At Shackleton, astronauts will not be able to use *Altair*'s shadow to judge distance, as it will be too far to the side. Another complication is the large descent stage under the crew, Hansen says. "With that big platform under us, you really can't see that well. *Apollo* couldn't see that well, either, but they didn't have this big front porch."

To that end, engineers are setting up various window placements and then flying simulations. Although this process can be done analytically on a model, researchers have found through experience that the human eye and human reflexes are best put into the design mix early, Hansen says. "It's very different—dynamically—having someone looking out the window and flying it."

Because of the visibility limitations of both *Altair* and the lunar environment, designers envision some form of augmented hazard detection. Possibilities range from floodlights to infrared cameras to scanning LIDAR laser systems. Ultimately, *Altair* will require a completely automated hazard detection system for the unpiloted cargo lander variants. Once the first landers have blazed the trail, electronic landing beacons or visual cues will be deployed to aid future flights.

Bruce McCandless, veteran of two shuttle flights, plays out the scenarios. "A simple approach is that you give the people already at the outpost some of these cans of orange highway paint. Now the problem is that you can't see just below you [from the *Altair* deck]. What you really need to do is sneak up on this thing and then let the automatic guidance take over in the end. You might use an electronic system or something like a GPS system around the Moon, but even now, the inertial guidance systems are up to the task. If you have an *Altair* landing and you look out the window and, lo and behold, the X on your heads-up display happens to be on top of this big international X someone has painted, you've made it. But if you're not, you run your track ball over to it to guide the craft to the right spot."

Just what will those flights look like? Current plans call for *Altair 1* to be a propulsion test on Ares 5Y, which is also the first test flight for *Ares V*. The unmanned mission will go into low Earth orbit. Flight designers are considering doing a trans-lunar burn, or perhaps a simulated burn long enough to get to the Moon without actually going. Hansen sees the first flight of *Ares V* as a golden opportunity. "To do a meaningful test of Ares 5Y, you need to at least send a mass equivalent to *Altair*, so why not get some good data?"

Altair 2 will actually have a more ambitious plan than the early *Apollo*s. The unpiloted craft will either touch down on the lunar surface or demonstrate an abort to lunar orbit, simulating a flight that is abandoned during an emergency on the way down. Either way, the second *Altair* will achieve lunar orbit.

Altair 3 will be the equivalent of *Apollo 11*, staging a landing of a crew on the Moon. *Altair 3* is designated HLR, "Human Lunar Return."

Plans are in flux for *Altair*'s design. Several industry partners have been tasked with evaluating the overall design concepts and safety of *Altair*. These companies are Andrews Space of Seattle, The Boeing Co. of Houston, Lockheed Martin Space Systems Company of Denver, Northrop Grumman Corporation of El Segundo, Calif., and Odyssey Space Research of Houston.

The New NASA: *Altair*'s Alternative Approach

Marcia Ivins is head of the Exploration Branch of NASA's Astronaut office at the Johnson Space Center. She has flown on five shuttle missions, spending over 1,318 hours in space. Her flights included work on both the Russian space station Mir and the International Space Station. Her many years of experience have given her insights into the way individuals and organizations contribute to a large-scale project such as Constellation.

We [at NASA] are an organization that is decades of tradition unimpeded by progress or lessons learned. We put together the *Orion* project and Ares and all these other things, and they are classically formed programs/projects/hierarchies of the way you do things. One of the unfortunate byproducts from a couple administrators ago was to remove the technical competence of the civil service agency and hand it to the contractor. So here, it used to be that civil servants actually built things. Their hands were dirty. They understood the mechanics. We had shops here and we built things here. Over the period of about fifteen years, that was eradicated. The effect of that has been that nobody in this agency in the past thirty years has built anything. Nobody in the contractor world has built anything for manned spaceflight. The shuttle was built in the mid-seventies. People have managed it, they've maintained it, they've fiddled around with the paperwork for it, but they have never actually built anything for it, particularly here at JSC. So

when they formed Laurie's [Hansen, manager of *Altair*] lander project, the thought was, *Let us form this as a small, skunk-works kind of a thing where you are exempt from the program process that is imposed on the other projects. You get a pass.* So Laurie's group actually works above the radar but underneath the process line, the intent being, can we regrow—in this agency—the capability to actually build something. The thought is that when the day comes, we can actually do the design in-house so that what we hand to the contractor is a build-to-print, rather than a set of requirements where they can charge us for whatever we didn't think of. So she's the rogue organization out there. We're sort of an experiment in progress here, in the way Laurie does business, and the way Constellation program does business. We [astronauts], as the crew, cross all borders and boundaries. We are the ones—as we have been historically—to cross every line and do much of the program's integration. We sit in the trench. We sit on the program boards. We sit in Laurie's group and Chris's (Culbert, NASA's Lunar Systems Project) group. We've got a finger in everybody's pie, and we become almost the only organization that integrates, to say 'do you know what they're doing? Do you know that that's not going to be convenient?' We become the connective tissue, and that is the role the crew has always served, because at the end of the day, we're stuck with whatever 'tissue' you put together. We *hope* it connects, because if not, it's us dangling on the end of it. So we have a very vested interest in connecting the tissue.

Studies are currently under way. *Altair*'s schedule is also in a dynamic phase, as it is dependent on its "mother ship," *Orion*. *Orion* still awaits its maiden flight, half a decade hence. But manufacturing of final flight hardware for the new *Altair* Moon ship may begin as early as 2015.

With the successful launching of the new Ares booster family, and with shakedowns of *Orion* and *Altair*, the first Moon mission is now slated for 2020. For the first time in half a century, humans will break the bonds of Earth's gravity and venture across translunar space. To NASA's Bret Drake, Chief Architect for Systems Engineering and Integration on Constellation, reestablishing that exploration capability beyond low Earth orbit is the priority. "Getting beyond Earth and to the Moon is a real big first step. It shows that we're serious about it and we're making great strides. [A lunar landing] is nearly twelve years away, so there's a lot of work between now and then, but that will be a good first step toward great new endeavors."

But returning to the Moon is only half the battle in creating a permanent human presence. To stay, we must build infrastructure, setting down a permanent outpost with power, communications, and transport. How—and where—will it all come together?

Chapter Three

Shackleton, the Home Site

Battered desolation spreads below us. The rugged landscape is a study in contradiction. It lies virtually unchanged from its formation nearly 4 billion years ago, and yet its dramatic face has been bruised by a hail of mountain-sized meteors, washed in torrential storms of solar radiation, baked by relentless heat, and freeze-dried in the wintry lunar vacuum. The Sun stays low in the sky, grazing mountain and crater rim with blistering heat. But in the ebony shadows, cheek by jowl with baking rock and sand, temperatures drop to −387° F.

We soar over endless rows of craters, large and small, some draped across flanks of ancient volcanoes, others resting uncomfortably on mountain peaks. Suddenly, a glint of sunlight breaks the unrelenting wilderness. It flashes on the edge of a black abyss. We have arrived at Shackleton International Outpost.

As we descend past 10,000 feet, solar panels come into view, blue jewels on a gray blanket. Their glistening wings carry power to a series of low dirt mounds. At 6,000 feet, the mounds resolve themselves into buried habitats, rigid cylindrical modules, and tuna-can-shaped structures buried under a meter of protective lunar regolith. We pass over a mining operation. A robotic rover lumbers across a series of freshly dug furrows. With less than 3,000 feet between us and landfall, we drift north, away from the habitation modules and into the landing zone. Earth is now hugging the horizon, dropping fast. Our engines kick up a fog of powder. A gentle bump announces a safe touchdown. The billowing dust drops instantly to the ground upon engine shutdown.

Establishing the first foothold on another world is a daunting task. Like nineteenth-century Arctic explorers, lunar architects face challenges of how to lay supply lines, set up living and working areas, provide power and communications, and build transportation infrastructure, not only from terra firma to the Moon but across the lunar landscape. And the first order of business is where to put it. The location must have access to nearly constant sunlight for power, and it must be in a spot visible to Earth almost continuously for safe communication. Does such a place exist?

NASA is considering several areas on the lunar surface for a permanent outpost, but one has been selected as a baseline. Shackleton Crater—at the lunar south pole—provides a realistic framework within which engineers and designers can study various architectures. The second candidate area includes a series of hilltops along the rim of Peary Crater, a great arena 73 km (45 miles) in diameter. At both sites, preliminary research indicates that raised crater rims and hills allow for nearly uninterrupted solar power at these sites, even in the dead of winter when the Sun lies lowest in the sky. Although Shackleton may not end up as the final site selection, it serves to inform real engineering and strategies so that humanity can move forward toward setting up the first off Earth permanent presence.

A sortie variant of an Altair *lander on final approach to Shackleton base. A Chariot rover ascends a slope in front of the primary outpost, which consists of rigid, inflatable, and small habitats. Three hab-carrying ATHLETEs are stationed farther out, and beyond them lies a field of solar collectors. Structures of an observation area glisten in the distance. The crater to the northeast (right horizon) is Mawson. (Art by author)*

M. Carroll, *The Seventh Landing*, DOI 10.1007/978-0-387-93881-3_3,
© Springer Science+Business Media, LLC 2009

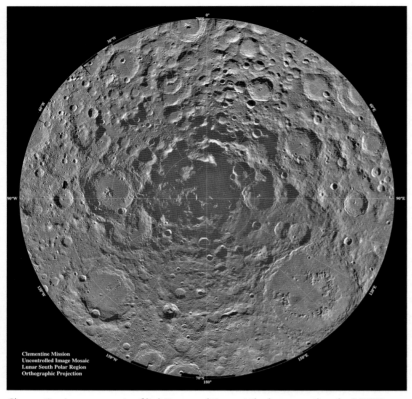

Clementine image mosaic of lighting conditions at the lunar south pole. (NASA)

Constellation designers and others believe the strategy of setting up a durable outpost on the Moon is a financially and logistically sound one, and one that hearkens back to the early days of Antarctic exploration. Planetary scientist Ben Bussey of Johns Hopkins University's Applied Physics Laboratory has twice explored the southern continent. He likens the opportunities of a lunar outpost to those afforded by McMurdo Station in Antarctica. "If all the science on the Antarctic continent had to be done carrying everything from New Zealand, no one would get much science done. But because we have McMurdo as a logistical base, expeditions can stage from there and do a lot more. Similarly, the outpost on the Moon will be a stepping stone into the Solar System."

At first blush, Shackleton Crater might seem an unlikely place to establish a beachhead to the planets. It's a place of stunning bleakness, a pockmarked landscape in eternal dusk. But to lunar base architects, it's the perfect site for a permanent outpost.

To paraphrase a British favorite, the Sun never sets on the rim of Shackleton. A vast amphitheater 19 km across, the crater sprawls like a bulls-eye across the

Where We've Been Before

Looking up at the Moon in the evening, it's possible to make out all the places where *Apollo* astronauts have explored. Mare Imbrium spreads cross the left half. The edge of this massive oval, the one that forms the Man-in-the-Moon's right eye, traces an arc ending at the 5:00 o'clock position. Within the bright highlands of the southern rim lies Fra Mauro, site of *Apollo 14*'s landing. To its left, out in the dark plains south of the Imbrium "eye," *Apollo 12* set down within 600 feet of the *Surveyor 3* spacecraft on rolling plains. Continuing counter-clockwise along Imbrium's border, we arrive at *Apollo 15*'s target, the Apennine Mountains, at about 2:00 o'clock. Travel down to the base of the bright nose, and we see *Apollo 16*'s landing site, hidden among the bright splotches of lunar mountains.

Across the bright highlands that form the Moon's nose, we travel east to the uppermost of three linked dark ovals, Mare Serenitatis. On its southeast edge, where the upper oval meets the middle one, *Apollo 17* explored the lunar highlands of Taurus Littrow. If we move straight down to the 7:00 o'clock edge of the center oval—Mare Tranquilitatis—we see the site of history's first lunar landing, *Apollo 11*. Seeing the location of the first lunar outpost may not be so easy. Located at the very edge one of the lunar poles, the base will bob in and out of view as the Moon wobbles in its orbit. Although the Moon keeps the same face toward Earth at all times, its orbit is not quite circular, so that it appears to sway from side to side. It also circles our planet on a slightly inclined path, so that its poles appear to bob toward and away from us. This movement is called *libration*.

Moon's south pole. From this location, the Sun seldom dips below the horizon. Over the leisurely course of a lunar day—lasting some 28 Earth days—the Sun bobs along the horizon, peering over the rolling hills of the Moon's southern highlands. Shackleton's raised rim assures almost uninterrupted contact with Earth, and a near-constant flow of solar energy. NASA's James Garvin comments, "As for the poles, there are regions of nearly continuous (albeit low angle) sunlight, well suited for solar power at the 10's of kilowatt level we need for human exploration."

The Moon's axis is tilted 5 degrees off the Earth-Moon line of sight. During half of each lunar day, Earth appears to float 5 degrees above the horizon, but for the other half day, our world disappears some 5 degrees below. However, geostationary satellites orbit far enough above Earth for many to extend the duration of a direct link with the lunar inhabitants during these periods. Relay stations at lower latitudes on the Moon have also been suggested as a means for the Shackleton encampment to keep in constant contact with terra firma.

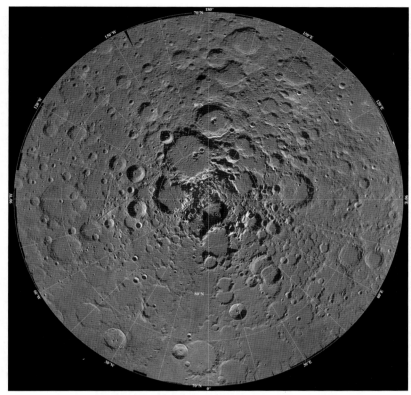

Clementine image mosaic of lighting conditions at the lunar north pole. (NASA)

Another aspect of Shackleton's environs appeals to lunar outpost planners. Simply put, the place may hold water. This fact first came to light when the *Clementine* lunar orbiter, a Pentagon-funded spacecraft, set its sensitive instruments to the task of creating the first detailed mineral maps of our nearest cosmic neighbor. Researchers managed to get even more science out of the small satellite. By beaming radio waves from *Clementine*'s communications antenna onto the lunar surface, radar experts on Earth were able to probe the subsurface of the Moon's soil. Underground, the radio waves bounced off of materials that mimicked ices. Ice usually sublimates—turns from ice into vapor—in a vacuum. Any Sunlight would cook ice from the exposed lunar soil quickly. But several deep craters at the south pole never see sunlight. These spots of eternal night act as cold traps and may preserve water ice beneath the surface. The ice is likely buried beneath a layer of regolith, protecting it from sublimating into the vacuum of space. The deepest, darkest craters and valleys at the poles are precisely the areas in which *Clementine* detected ice-like radio reflections.

This remarkable, though indirect, discovery was bolstered in 1998 with the arrival of NASA's *Lunar Prospector* orbiter. The craft charted hydrogen escaping from both poles. The most readily available source of hydrogen in the lunar environment is thought to be water ice.

NASA Goddard's James Garvin says, "The important sidebar to all this comes when you think about collisions and the bringing of volatile materials—water-ice, methane, clathrates, whatever squishy stuff you like—and depositing them on the Moon. For the last 2 billion years, the Moon has been in its resonant lock with Earth,[16] and the cometary volatiles that have survived in the lunar system have been deposited at the poles. It's inescapable. So what happened to them? Did they sublime away? Did they become sequestered? That's the issue that's been contentious in the planetary science community—about whether storage of redeposited ices in lunar soils as small bits or hockey rinks—is possible." In other words, although the jury is still out, the Moon's south pole may harbor ices from primordial epochs.

Water is a valuable resource for permanent lunar settlers. Aside from filling a drinking glass, water can be split into hydrogen for fuel and oxygen for breathing. When reversed, this splitting process (called electrolysis) releases electricity. (A similar process is used in fuel cells aboard many spacecraft, including the shuttles, for power.) The south pole's Aitken impact basin, with Shackleton at its edge, may be a vast repository for this treasured resource. Lunar geologist Paul Spudis, says, "Such a site can potentially be an 'oasis' in the lunar desert." It is water, not gold, that lunar explorers will covet.

Water is not the only resource available to astronauts. The Moon's sterile soil, called regolith, contains valuable materials for making a permanent home. At NASA's Glenn Research Center in Cleveland, Ohio, engineers are designing ways to coax oxygen from the dirt. The key, says Glenn's John Caruso, is automated rover technology. "Believe it or not, there's a significant supply of oxygen in the regolith, roughly 40 percent." NASA's Chief Scientist Jim Garvin adds: "We know there is plenty of O_2 in the lunar soils, concentrated in materials from which it is relatively easy to extract (in minerals such as ilmenite). Some lunar soils (including samples from *Apollo17*) are 10 to 13 percent ilmenite by volume, making them good sources for extractable O_2."

But Caruso warns, "It isn't easy to get out. Still, it's not nearly as difficult as it is to haul up ice from the bottom of a cold crater. We take a supply of hydrogen, run it through a batch process, and collect the oxygen via attaching it to the hydrogen. Then we'd separate the hydrogen and oxygen, save the oxygen in a gaseous form, and recycle the hydrogen back into the system. It's a single supply of hydrogen that can be continuously used to harvest the oxygen."

Caruso is project lead for NASA Glenn's In Situ Resource Utilization (ISRU) project. His team is tasked with developing ways to manufacture oxygen and other resources from lunar material on site—*in situ*—at the outpost. Caruso envisions two main types of rovers: a small prospector would seek resource sites and drill or dig, while a larger excavating rover would carry the material to a processing plant. Clever designs enable the rovers to operate on

16. In other words, the Moon keeps the same face toward Earth, turning once for each time it circles Earth. This is true of most moons in the Solar System.

All in a Name

Features at the lunar poles, where sites for permanent outposts are under consideration, are named after Earth's great Arctic explorers. The two primary sites to date are the rim of Shackleton Crater in the south, and the Peary Crater region in the north. The mapmakers have given noble names to these craters.

In 1914, Ernest Shackleton commanded the Imperial Trans-Antarctic expedition. Seven years earlier, Shackleton had led the British Antarctic Expedition (which failed to reach the South Pole), but the Trans-Antarctic voyage was more ambitious. The South Pole had already been attained by Roald Amundsen in December of 1911. Shackleton wanted more. He planned to cross the Antarctic continent on foot. Shackleton's ship—the *Endurance*—was to make landfall in the vicinity of Vahsel Bay, while another ship landed on the opposite side of the continent at the Ross Ice Shelf. The second ship was to deposit a series of caches across the Ross Ice Shelf leading to the base of Beardmore Glacier, where Shackleton anticipated emerging after the long trek. The ship *Aurora* made its way to the Ross Ice Shelf, and despite hardships and the loss of three crew members, the expedition set down the series of critical supplies. But Shackleton's team never found the precious rations. Their ship *Endurance* became icebound, eventually splintering under the pressure of the ice and sinking. Under the most brutal of conditions, Shackleton led his

27 men on a series of journeys across 200 miles of oceanic ice flows, rugged mountains, and treacherous glaciers. The crew used the lifeboats and material from the ill-fated *Endurance* as sea vessels as well as shelters and sledges. Shackleton finally set sail across the stormy ocean on a tiny improvised craft. The desperate sea voyage paid off: after 22 months in the Antarctic wilderness, all expedition members were rescued.

Robert Peary led a party of five explorers to reach the vicinity of the North Pole in 1909. Whether Peary actually reached the geographic pole is controversial, but Peary was an accomplished polar explorer in his own right. Unlike other explorers, Peary adopted strategies of the Inuits, traveling by dogsled and wearing clothing styles that native peoples of the Actic had been perfecting for centuries. Peary even built igloos while on the expedition. His voyages across Greenland located the northernmost land on the planet. Peary charted several important landmarks and did pioneering work in the documenting of magnetic variations.

The techniques used by both Shackleton and Peary helped shape the strategies of Arctic explorers for decades to follow. Lunar architects suggest that some of their strategies, like caching supplies and using redundant vessels, may well be applicable to future Moon and Mars expeditions.

less than 100 watts of power. Vehicles that must work in cold environments such as crater floors may need a nuclear power source, and a new one is coming on line. Called a Stirling engine, the power is generated by alternating hot and cold cycles, Caruso says. "We use a small radioisotope source to make it hot, and the other side is cold. It's a bit like a piston in your car, although your car's piston has a lot more parts. The Stirling engine is very simple." Engineers are also looking at batteries recharged by photovoltaics (solar cells) or fuel cells similar to the power plants used on the space shuttle today.

The excavating rover can be battery-powered. Each time it drops off a load of regolith at the solar-powered processing station, the rover recharges from power at the station through the use of an inductive recharger. "It's just like you have on an electric toothbrush," Caruso says. "You can recharge by getting up against it" without plugging any hardware in.

With a small armada of these rovers, oxygen for the Shackleton outpost may eventually be produced on site. But the key word is "eventually," says Chris Culbert, manager of NASA's Lunar Systems Project. "It's a complex prospect. You're talking about some kind of automated system that will crawl into those environments and somehow find a resource for you and dig it up and produce it and send it back to you. That gets very complicated. It's not the kind of thing you want to bite off at the beginning. There are enough other tough problems to solve first."

SITE SELECTION

The rim of Shackleton crater is fairly rounded. Several sites along its elevated ring offer prime real estate for a permanent home. As any good realtor will tell you, it's all about *location, location, location*. With Shackleton's potential for local water, oxygen resources, and clear views of Sun and Earth the area seems to have what it takes to establish humankind's first permanent presence on the Moon.

Temperature is another consideration in site selection. In the equatorial *Apollo* sites, daytime temperatures reach 100° C, while night temperatures dive to −150° C. But because of the Sun's apparent path across the polar skies, the poles offer dampened temperatures that average approximately 50° C below zero. Coincidentally, this is the average temperature of the Martian equatorial regions.

Creating and sustaining a permanent human presence on another world will require one of the greatest efforts in human history. For the first time since *Apollo*, nearly every major technological institution in the United States is involved in one project, and other countries are becoming an integral part of the project. NASA's Chris Culbert knows this well. "There is a wide range of expertise you need to put together a decent architecture like this. We're studying all the systems you need to create a lunar architecture: structures, robotics, avionics, software, human habitation, life support systems, communications, and power." While financially incremental, the complexity and scope of the effort is unparalleled.

Once a beachhead is established on the lunar surface, the world's scientific community—as well as its entrepreneurs—will have unprecedented access to the new world. But pitching camp at Shackleton will be no easy task. It will require a lot of assembly, and much of it must be done by humans with hands-on experience. The construction of an outpost calls for hard physical work, skill, and mobility. Hours spent in the lunar environment call for new spacesuits with advanced architectures, and designers have made a good start.

ADVANCED SPACESUITS

The space environment is completely hostile to terrestrial life. The only thing standing between the deadly lunar environment and an astronaut will be a spacesuit. A suit must provide pressurized air to breathe, communications, and thermal control to protect its wearer from the soaring temperatures in direct sunlight and the blistering cold of shadowed areas.

Operating on the Moon, *Apollo* astronauts faced two major challenges: balance and mobility. To Earth viewers, balance was the most obvious problem. The portable life support system (PLSS) backpacks worn by the dusty

Comparison of the profile of an Apollo *lunar suit (left) with the new Mark III (right). Note how far back the life support pack leans on the* Apollo *suit. The Mark III has a smaller pack and a center of gravity that helps to keep astronauts balanced.*

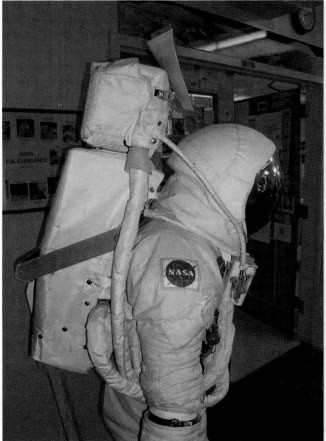
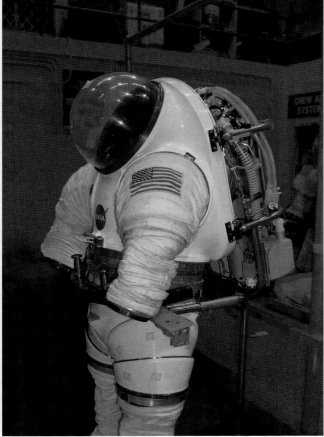

dozen tended to pull the astronauts backwards, forcing them to lean forward for balance. The famous "Moon Hop" adopted by lunar explorers was a necessity: lean forward or go down into the dust.

The next generation suits will have a center of gravity farther forward than earlier suits. Pivotal to the new suit studies is the Mark III. "Mark III is the fundamental architecture that they've been working from," says suit design veteran Joe Kosmo. "It has a lot of the advantages of a planetary suit; it's robust, has mobility, and has longevity. We've been working with this suit for over twenty years. It started off being a higher-pressure suit for the space station so we could eliminate pre-breathing. But as it turned out, it became a good testbed for some of the necessary planetary surface operations."

The Mark III has dramatically increased mobility over earlier suits. Engineers have replaced cables with high-strength webbings to distribute loads induced by pressure and human-induced stresses such as bending and twisting. By geometrically tailoring the mobility joint systems and incorporating specific patterns in the suit's form, designers make the aspect of "ballooning" work for them rather than against them. Overlapping joints, arranged much like a tomato worm or insect carapace, afford strength as well as flexibility. It's a difficult assignment. Many of the *Apollo* moonwalkers ended EVAs with bruised and bleeding fingernails from long periods working in the stiff gloves. Constellation's Glen Lutz says, "If suits are a technology, gloves are an art form." Lutz wants future explorers to have easy access to the lunar wilderness. "We've come up with something we call the work efficiency index. We want to be able to jump into the suit and go out quickly, so we can maximize our time out the door vs. the prep time. The suit is a

Apollo *gloves like this one were stiffened by cables and layers of material. (photo by author)*

The venerable Russian Orlon ("Eagle") environment suit is accessed through a rear port behind the backpack. This workhorse is worn by both Russian and American astronauts aboard the ISS.

The Orlon has its heritage rooted in the suit designed for the first Russian moonwalker. The lunar suit was called Kretchet (Gyrfalcon). Both suits were the first to use a metallic, rigid torso, and both use a rear entry port. (Ben Guenther/Karl Dodenhoff/myspacemuseum)

critical piece of life support, but the crew doesn't need to think about it in that way. They need to think about it as a pair of overalls so they can go out and do their job."

The Mark III sports a slimmer life support backpack built into the suit. The cryogenic gas (liquid air) in the current design weighs 12 lbs and is good for a period of an hour, but design baseline will put the backpack duration at a nominal 8 hours, with emergency supplies lasting considerably longer. In addition to oxygen, the pack contains coolant and a CO_2 scrubber.

Donning the Mark III is similar to putting on a Russian Orlon spacesuit: the backpack serves as a rear entrance. Astronauts step into the suit through the "hatch," then close the backpack behind them. The Mark III incorporates a rigid torso and flexible arms and legs. The helmet is attached to the torso section but features a bubble-like visor for a wide field of view.

Light levels vary dramatically at the lunar pole. Like the *Apollo* and shuttle/ISS suits, the helmet

Comparison of the Apollo *lunar surface suit (l), the shuttle/ISS EMU suit, and the Mark III. The Mark III advanced spacesuit provides a springboard for future lunar spacesuits.*

has a set of sunshades that can be folded away in low light levels. Lunar suits will also have cameras and lights similar to the more advanced shuttle/ISS helmets, enabling astronauts to work in a wide range of lighting conditions. Off-the-shelf LEDs have been tested under rugged desert conditions at night, with good results.

Suits will be modularized, making for easy replacement of a variety of hardware, from dust-saturated connecting rings to delicate components of the environment pack. And the Moon's talcum-like dust is the one problem cited most often by veteran *Apollo* astronauts. Astronaut John Young, spacecraft commander of *Apollo 16* (April 16–27, 1972), said, "On the Moon, what hampered us the most was the dust getting in our wrist rings and our neck rings. I don't think we could have done a fourth EVA because it was getting so bad. When we go back it'll be the same thing."

Designers are attacking the problem of lunar dust on two levels. The first order of protection is to exclude the dust from the interior of the suit itself. Engineers have developed enhanced outer garment materials, tight weaved Teflon that is heat-sealed. The material must enable easy dust removal; the ideal is that it doesn't attract dust in the first place. The second order of protection involves dust exclusion seals in the bearings. "We've found that felt with lubricant provides good protection," according to Kosmo. "So far, our felt is not dense enough, but we're working on that, and it looks like the concept will work. We've pretty much solved the problem from the hardware standpoint."

Moon grit in living areas constitutes a second problem. Strategists are exploring two alternatives. The first is to treat airlocks as a sort of mudroom. The airlocks would be equipped with air showers or mechanical mechanisms to brush dust from the suits. But recently, a second alternative has been attracting more interest. If an astronaut can step through a backpack port directly into a self-contained suit, why not leave the suit outside?

Walk a Mile in My Suit

"You can't really understand a spacesuit until you get inside one," says Johnson Space Center designer Joe Kosmo. Kosmo should know: he has designed suits for every human space mission from project Mercury to the ISS. "I was 21 when I got the call from the Lewis Research Center. They said, 'We're going to be working on life support systems and spacesuits, whatever those are.' And I said, 'It sounds interesting, but I don't know anything about spacesuits.' So the guy at Lewis said, 'Well, nobody else does either.' Quite a bit has changed."

As an example, Kosmo pulls an old *Apollo*-era glove from a locker. The glove was worn by Deke Slayton during the 1975 *Apollo/Soyuz* Test Project flight (the first Soviet/U. S. joint flight). "They were custom-molded to the individual's hand. You can see there wasn't a lot of mobility designed into this glove. Once you pressurized this glove, it was all you could do to grab something. Ken Mattingly wanted his glove molded into a position like a claw. You fit the crewman with what he needs, and if he's happy, you're happy. The glove had a cable system housed in a yolk arrangement anchored at the wrist. It took quite a bit of force to work the glove. At the time it was the leading edge of our technology. There were convolutes at the joints, too. Once you pressurize the suit, it wants to expand, and the cables carry the load. The shoulder also had a cable transfer system. It didn't have a lot of mobility and took a lot of force to move. But the shelf life is only about four years. We've advanced since then, not only in the gloves but in the entire suit. For going back to the Moon, we're talking about permanent presence. You want something that's going to last a long time: robust, very durable, and designed for lunar surface operations. We got an insight when we developed the shuttle suit. One thing we learned from *Apollo* is that you don't necessarily want to make one suit try to do two jobs (an intravehicular suit and an extravehicular suit), because you have to compromise on both ends to make a happy medium. We also had a lot of potential single point failures on the *Apollo* suit. There are all these external connectors that had to be mated; the crewman had to make all these connections every time he went out of the vehicle. There is wear and tear on the hardware. You also had the problem of dust. We did not provide real dust protection. I'm amazed it held up as well as it did. I'm pretty impressed with what it did in the circumstances we subjected it to. But we've learned a lot, and we're applying those lessons to the next generation planetary exploration suit. We've come a significant way from where we were with *Apollo*. We've developed different kinds of mobility joint systems, flat-pattern joints, tuck-fabric, rolling convolute joints, things that will really enhance mobility. So instead of a balloon, you are left with something more like a tomato worm. We may not have reached an optimum, but I don't know if you ever really can reach an optimum. You have to come up with an acceptable design that satisfies the needs, and I think we're getting close to that. Having been in the business long enough and looking ahead for our needs—projecting what's going to occur—and knowing the history of the suits we've been working with, I've got a lot of confidence in going with the hardware we're focusing on now."

The concept is called the "suit port." Imagine a pressurized rover with spacesuits mounted on the outside. The backs of the suits are sealed against a small doorway on the side of the vehicle. Whenever an EVA is called for, crew members simply step from the interior of the rover directly into their externally mounted suits, seal the door behind them, and walk away. Because of the hostile lunar environment, suits may be curtained by an awning that provides protection from dust and UV radiation. Kosmo likens the choice to "hanging your clothes in a closet as opposed to just throwing them on the floor."

The idea of suits taking the place of an airlock may be carried even farther. In the early days of outpost development, habitats may be lofted to the Moon without heavy airlocks installed. Instead, the small pressurized rovers can be docked to the side of a hab and serve the role of an airlock. The advantage is that this airlock would be completely portable, able to plug into any point on the outpost. *Apollo 14* LM pilot Ed Mitchell likes the concept. "Getting in and out is a significant obstacle. Maintaining that airlock or transition chamber is a big deal: if it's very big and you pressurize and depressurize, you're wasting an awful lot of air. You want to keep [airlocks] fairly small. On the other hand, having an airlock mechanism where you can leave the dusty pressure suits in a quasi-external environment sounds like a really good idea."

Visors or optic surfaces may also need advanced coatings or coverings to resist scratching. Designers are even considering strip-off visors akin to those in use by motorcycle riders. A scuff-proof coating of polysulfone material can also be applied to visors. Shuttle suits now use anodized silver on the visor and impact-resistant polycarbon on protective visors and helmets. Low emittance protective UV visors are used for shadow, gold or silver for direct sunlight. It's a proven material that has worked well on *Apollo* and on the shuttle/ISS.

In its current form, the Mark III does have its drawbacks. The most critical is weight. But Joe Kosmo points out that with further research, lighter composites can be developed. "[Critics] keep coming back to 'it's too heavy,' but astronauts who have tried it on in 1/3 and 1/6th g on the KC 135 [NASA's low-gravity aircraft simulator] disagree. They've actually done handstands in it."

A second potential challenge comes with the Mark III's bulkiness. Although a distinct improvement over earlier suits, the Mark III would hog precious elbowroom aboard

Astronauts stored their two grimy moonsuits aboard the Eagle *after history's first moonwalk. Constellation designers hope to avoid such messy scenes with the use of externally mounted EVA suits. (courtesy NASA)*

Orion, where space is at a premium. Some budget-conscious managers would like to see a single suit used for both intravehicular (inside the spacecraft) and extravehicular (outside) activities. Heavily modularized suits would enable the flight crew to change out elements for different applications.

"For Constellation, we're trying to put all that into one system," explains Glen Lutz. "But we've got almost conflicting requirements: two configurations and two roles for one suit. Suits would share lower arms and legs, as well as helmets." For surface EVAs the crew would add a different torso section and a TMG (thermal micrometeorite garment), an outer shield of layered ortho fabric, Kevlar, silver mylar, and other materials for micrometeroid protection. Contractors have yet to announce final materials. The suit would have to be able to serve an 8-hour EVA every other day for six months and withstand up to one hundred sorties. "That's why we're after more mobility, better comfort, robustness, and donability [ease of putting it on]," Kosmo says.

One system due for an overhaul is the familiar snoopy-hat communications system worn by *Apollo* astronauts. The cloth cap held communications microphones and earphones against the astronauts' heads. For the next generation system, designers hope to have an integrated communications system married into the suit. Custom-fit hardware drives the cost of the program and also drives the logistics and support of those individual sizes. Smaller- or larger-than-average sizes cannot be used interchangeably, driving up costs. An integrated system is less costly and simpler than providing custom-fit communication carrier assemblies, soft goods that are difficult to keep clean in the gritty lunar environment. The snoopy hats also proved vulnerable to handling. Small microphone booms and pigtail-lead wires

Both the space shuttle/ISS and Apollo *suits (left) feature helmets with retractable sunshades and visors. The Mark III (right) will have a similar arrangement but will use materials resistant to the abrasive lunar grit. (photos by author)*

were subject to handling damage over a period of time. On shuttle and *Apollo* flights, microphone booms suffered flex-failure when they were bent to fit the crewman's lip area. The "pigtail" wire bundles also saw damage from routine handling as they were crimped or pinched. An integrated system will avoid the stresses of periodic connecting and disconnecting, and the built-in nature of the equipment will circumvent the complexities and cost of custom fitting.

Some analysts disagree with the single suit approach. They advocate a lightweight flight suit similar to the ones worn by shuttle astronauts while inside the craft, and a separate suit for EVA work (comparable to the shuttle's EMU). Joe Kosmo argues that, "We learned our lessons from *Gemini* and *Apollo* and applied them to shuttle. The shuttle has two suits, an intravehicular and an extravehicular. If you try to combine them, there are just too many compromises you have to make."

At this stage, Constellation architects are putting everything on the table, examining every possibility and suit iteration in a creative design environment. They've even looked at alternative suit designs. One such alternative is called a mechanical counter-pressure suit, also known as a skin- or bio-suit. Instead of putting an astronaut into a suit that amounts to a pressurized balloon, mechanical counter-pressure suits form a supportive skin against the body. A thin and flexible thermal covering would keep body temperatures equalized. Advocates believe this approach will solve many mobility problems.

Many planners assert that the mechanical counter-pressure suit offers an interesting concept but does not yet adequately compensate for body physiology. The amount of counter-pressure required for protection against a vacuum necessitates complex—and expensive—multilayers. Another challenge faced by the skin-suit proponents is developing an equilibrium on the outer layer that will apply support over all parts of the body. When an astronaut bends an arm, for example, a void opens up at the elbow or under the arm. The problem: how to apply constant pressure there. Compensating pads have been embedded at those sites to distribute the loads, but they are uncomfortable and nearly as constraining as cable or web systems. Thoracic pressure is also a problem. Subjects may experience difficulty breathing with increased pressure on the chest and abdomen. But a skin suit is not the only thing that's tight these days. Money is, too. With current budgets, most engineers see mechanical counter-pressure suits as a promising technology still in its infancy. Still, the Constellation teams continue to look to industry. Kosmo calls the process "a cross-fertilization of information. In the early days, we worked closely with the air force on anthropometry—human sizing—at Wright Patterson [Air Force Base]. Because of the lack of technology money we do look around to see what other people are doing and adapt some of the aspects of other work to our needs. We don't have the kind of money that we did in *Apollo*, and we probably don't need that kind of money to get the job done. It's more application of what we've learned."

No matter how good the suits are, there are some places that humans simply cannot go with current technology. Although the average polar temperatures are a relatively balmy −50° C, it's a different story in the dark hollows of some permanently shadowed polar craters, where temperatures hover at a deadly −233° C. It is here that the robots must take over.

Chapter Four
Robot-Human Combo Systems

Two Chariot transport vehicles approach a newly arrived Altair cargo carrier, where an automated ATHLETE system prepares to offload a massive payload. In the distance, a sortie-configuration Altair awaits the return to Earth.

A flash of light races across the field of stars, announcing the arrival of another *Altair* cargo vessel. Stars are seldom seen on the Moon, but in the deep shadows of the south pole, Shackleton inhabitants see them frequently.

Altair lands near a homing beacon a safe distance away from the outpost. The twelve-wheeled Chariot vehicles lumber toward the landing site as a robot ATHLETE offloads the precious cargo—a module that will add

79

M. Carroll, *The Seventh Landing*, DOI 10.1007/978-0-387-93881-3_4,

nearly 20 percent living space and bring scientific laboratory facilities to the lunar settlement. A permanent international crew of seven already tends Shackleton base, with more to come.

NASA and other space agencies are exploring a balance between robotic and human-tended systems, including remotely controlled lunar rovers and automated robots programmed to assemble large structures such as habitats. *Apollo 12*'s Alan Bean puts it this way: "Humans have many good qualities, and one of them is adaptability and learning to adjust to the situation. They have other undesirable characteristics. We've got to have oxygen to breathe; we've got to have pressure in which to operate; we've got to be within a certain temperature range; we can only think so fast. So humans have limitations we have to deal with, but they also have tremendous advantages, as we've seen with the construction of the [ISS] and the repair of the Hubble Space Telescope. So the combination of the best automatic machinery you can build and human adaptability and flexibility make the best combination for exploration. There is no such thing as unmanned or non-human spaceflight; it just depends where you put them. If you've got a robot, you still have humans doing all this. They're just sitting in Mission Control somewhere in California or Houston. You're never going to do away with the human part."

The goal is not to displace human explorers, says the Lunar and Planetary Institute's David Kring, but to augment their work, do some operations more cheaply, and keep the humans safe. "You can spend a few hundred million dollars that will, in the end, save you billions just by building robotics."

One such robotic brainchild is called Chariot. The vehicle has a dozen wheels arranged in six sets of two, each independently steerable. Crab mode allows all wheels to turn in the same direction concurrently, giving the vehicle a turning radius of essentially zero. The craft can also steer around a point centered under one set of wheels, or even a point somewhere off in the near distance. Chariot can be remotely piloted but also has capacity for crews. The driver stands in a turret (gondola) at one end that spins 360°. Most test

The Dextre *robot, a Canadian-built "special purpose dextrous manipulator," moves across the Destiny Laboratory Module of the International Space Station (ISS), completing tasks prior to the deployment of Japan's Kibo pressurized science laboratory. Dextre has arms 3 meters in length and can attach power tools to its "hands." (courtesy NASA)*

drivers prefer to drive the rover with the turret in the back so they can observe all wheels. This arrangement gave rise to the vehicle's name, as it resembles a horse-drawn Roman chariot. Another configuration has the gondola side-mounted.

The Chariot's deck can be raised off the ground for driving, or lowered to ground level for offloading. This feature also enables the rover to bulldoze with a blade on one end. Several sets of rear foot restraints allow for standing passengers at the back. The craft consists of a hollow steel tube frame for easy attachment of pressurized cabins, habs, supplies, etc. No roll bars crown this Moon buggy, because engineers want the deck to be open for easy loading of payloads. If astronauts top the vehicle with a hab or other human-inhabited pressurized compartment, roll bars can be affixed directly to the payload itself. With a top speed of 20 km/hr, the Chariot

Someone stands in the gondola of the Chariot rover for scale. Note the LIDAR and other navigational systems in the front to help guide the Chariot rover remotely.

could become the adaptable Small Pressurized Rover under consideration for lunar surface operations and long range exploration. Safety-conscious mission planners envision dual-rover sorties, such that if one rover fails, the other can carry all crewmembers back to base.

In addition to sorties, rovers can fill in as the first cosmic pack mules. The vehicle's LIDAR (light detection and ranging) laser system enables Chariot to follow a human companion. Running in low gear, the rover can operate in a mode that senses a homing beacon on an astronaut, or it can simply watch the movement of the figure and follow at a specified distance behind. Prospectors of the old west would feel right at home.

Most important for the early establishment of an outpost, Chariot can do tele-operated construction and already has in field tests. The front of the craft houses laser guidance and stereo camera systems, giving it the flexibility needed for telerobotics. Designers envision habitats that have their own platform and leg structure. The habs would be picked up from a landing area, offloaded in a stowed position, perhaps by a remote crane, and would remain undeployed until the rover arrives at the outpost site. Chariot would move the hab into position, and the hab would lift itself up while the rover lowers itself to drive off. In this way, Chariot can autonomously prepare an area for a hab, carry the hab to it, and deploy it onto the site, all under human-tended remote control.

This somewhat daunting scenario has already been tried. Styrofoam and plywood habitat mockups have been fitted onto the Chariot testbed.

The aesthetics of hardware contributes to the quality of life for those who will be living and working in the harsh lunar environment. On this count, the advanced rover comes through. The Chariot is beautiful, with golden struts, glistening white housings, and graceful lines. Rob Ambrose, the project manager for lunar robotics, explains that this is an important part of the approach. "If you give them some creative leeway, they tend to stay late. It ends up not costing any more to make it attractive." But Chariot's grace is only skin deep: it has the heart of a bull. The combination of strength and low weight weaves a common thread through most lunar designs. Says NASA's Chris Culbert, "Here on Earth we just build them massive to make sure they're going to hold up. That's not such a hot approach when you're taking it to the Moon. 'Light' is very important to us. You have to find the right balance."

While Chariot toils in a gravel simulation yard at Houston's Johnson Space Center, engineers send a futuristic, crablike rover through its paces across sand dunes north of Baker, California. Known as ATHLETE, the spidery six-legged device is about the size of a mid-sized SUV. The craft's name stands for All-Terrain Hex-Legged Extra-Terrestrial Explorer. Each face of the rover's 4-meter diameter hexagonal core has a set of stereo cameras and laser rangefinders to navigate over multiple types of terrain. Each leg has a wheel so that the craft can be driven, but wheels can be locked and used as an anchor, or "foot." Each wheel, in turn, can be swapped out for a claw or power tool. The current test version weighs 850 kg and has a reach of some 6 meters, but the operational version will weigh in at 2,500 kg, with an impressive payload capacity of nearly 15,000 kg (about 15 metric tons) in lunar gravity. The behemoth's arms will have a reach of 8 meters. The long reach is important; it enables the ATHLETE to be secured to the deck of the *Altair* cargo lander, with the cargo on top of the ATHLETE itself. After landing, the ATHLETE can use its long arms to step off the lander deck, which is 6.5 meters high, carrying the cargo with it. Brian Wilcox has been developing the ATHLETE at NASA's Jet Propulsion Laboratory. He outlines the procedure: "The current concept is that a single six-legged ATHLETE would just walk off the top deck of *Altair* carrying the payload. The ATHLETE limbs would be attached to a pallet that has the payload affixed to it, and that pallet would have a separation interface (i.e., pyrotechnic charges that fire to release bolts) that will allow it to be freed from the top deck. The ATHLETE limbs would unfold from their stowed position and be able to reach all the way to the ground. Four of the limbs would stretch to the ground and roll while two of the limbs (call it the back) would step on corresponding 'hard points' on the *Altair* top deck. They would support some of the weight of the payload while the pallet shifts forward, changing the leg pose as the pallet shifts, so as to keep the two wheels stationary on the nodes of the tubular space-frame making up the *Altair* structure. Once the pallet has shifted far enough forward to expose two more hard-points on the *Altair* deck, the vehicle would stop rolling forwards long enough for the two back limbs to re-adjust their wheels onto the two new hard-points. Then the vehicle would roll forwards again, until the back limbs can step down onto the ground."

ATHLETE's design allows for other uses. The craft could also be topped by a pressurized crew compartment, serving as a human-carrying rover for construction or long distance science sorties. During such trips, Chariot-class rovers might transport crews away from the habitat-carrying ATHLETE for short excursions. ATHLETEs might also be stationed on high ground to act as solar-powered communications relay stations. The Chariot has a

The End of an Era

The first great era of human lunar exploration came to an end on December 14, 1972, as the lunar module Challenger *lifted off from the mountains of Taurus-Littrow. The crew of* Apollo 17, *along with its predecessors, left behind tons of equipment and artifacts at six different landing sites, including three rover "Moon buggies," twelve PLSS life-support backpacks, six U. S. flags, six LM descent stages, and 100,000 footprints. At every Apollo landing site, a suite of scientific instruments sprawls across the gray dust, paying silent tribute to thousands of scientists back home who continue to unravel lunar mysteries. In a shadow cast across history,* Apollo *bequeathed new scientific paradigms about the Moon and Earth, and the promise that one day, someone would be back.*

Harrison Schmitt was aboard the last crew to visit the Moon. He was also the first scientist there. Here, he shares some thoughts on rovers, exploration, and advice to the next generation of lunar inhabitants.

I never felt uncomfortable at all. Communications were great. The rover performed beautifully. There was no feeling in my mind that we were at any kind of risk beyond just being 250,000 miles from Earth. We could have walked or run back [to the Lunar Module from the rover] quite easily. I tried [to get NASA to be] more relaxed about walk-back distances. I'm certainly encouraging them to be with future plans. It partly depends on how well the new suit design turns out. They are currently in the procurement process to get a new lunar space suit, and I'm working hard to do better than we did with the *Apollo* suit, even though we did a great deal with the *Apollo* suit—the A7LB, as it was known—but clearly after forty years we ought to be able to improve on that. We need to particularly improve on the gloves. The gloves are the most debilitating and inefficient part of the suit.

As for returning to the Moon, we knew that it would be quite a while. I was surprised that it's taken this long,

and took several presidents before we finally decided that it was time to become spacefaring again. That is unfortunate, but you can't cry over spilt milk. You just keep working with the hand you are dealt.

We've just begun to discuss rover details in the NASA advisory council that I chair. There's been a great deal of work done at the JSC on potential new designs for rovers. My own preference would be to have a rover that comes in two phases. One would be an unpressurized rover that can be used in the early excursions around building an outpost, and then you would be able to outfit the unpressurized rover chassis with a pressurized cabin. I think a phased approach is the best planning approach right now. Improvements I'd like to see include on-board consumables, automatic position location, and continuous, high resolution, stereo imaging of exploration sites and sampling activity.

I would recommend that future crews get a lot of field training that is simulation-based. Beginning with *Apollo 13*, we had about a week each month of actual field training that was run as if it were a lunar mission. That was very important, both in becoming comfortable with the operations of exploration in space, and also in learning the geology that was relevant to what they might find in their particular site. You need to make sure that you have absolutely maximized your physical conditioning of your arms and hands. You need personal trainers so you can do it in a systematic way so that no one goes to the Moon better prepared, physically, than you are. I wish I had that kind of advice and council before I went. I don't care how good that suit's going to be. You're going to wear your arms out. Only your arms get tired. The 1/6 gravity makes it really easy to work, but you need to be conditioned superbly, as the Space Station EVA astronauts are today. You also need to be disciplined enough not to wear those muscles out early on. I don't think we were as well prepared physically as we should have been, nor were we disciplined enough at the start that we kept those muscles from fatigue.

The ATHLETE is designed to negotiate rugged terrain while carrying payloads or tools for construction. ATHLETES can also carry pressurized habitats. (Photo courtesy of NASA/JPL)

spring-damper suspension and so can go much faster (about 20 km/h), while the lunar version of ATHLETE is now planned to go only about 5 km/h."

ATHLETE can be commanded directly from suited EVA astronauts using a keypad that could be mounted on the arm. It can also be driven remotely by crews back in the shirtsleeve safety of the outpost using a laptop or console. ATHLETE can also be directed by controllers back at Houston or other Earth control centers. The vehicle's software derives from that developed for the highly successful Mars Exploration Rovers *Opportunity* and *Spirit*. Like the twin Mars rovers, engineers feed command sequences to ATHLETE, sending it to designated waypoints and executing automated sequences to perform activities at each waypoint. Rover designers are also equipping the vehicles to retrieve samples, guided by the astronaut shining a laser on the target.

Reinventing the Wheel

Although many hundreds of scientific and engineering papers were written during and shortly after the *Apollo* Project, much data was never documented for later generations. Now, designers are scrambling to rediscover what was done, how it was done, and details on structure, materials, etc. "It's always nice to have somebody who has done it before," says John Caruso. "Some of the technology is obsolete, but clearly there is a significant amount of data that will be of great help." Engineers at NASA Glenn are rebuilding a dozen *Apollo* lunar rover wheels. "We're going to do ground testing on them and compare them to data we have from the Moon and then we'll develop the next lunar wheel. Where we do have data from the Moon, that's pretty rare stuff. You want to take advantage of all that you can find and make use of."

Jeff Hanley, program manager for NASA's Constellation project, sees devices like Chariot and ATHLETE as critical. "Automation enables the human element to be productive. It frees up the human from having to do the mundane systems management. Automation will be most valuable on the surface of the Moon. We automate so the crew can be preparing for landing, Moon walks, field geology, or constructing an extension to the outpost, doing the things that humans are good at: in situ problem solving. Even on ISS, we spend too much time sustaining or maintaining systems. We want to free crews to do what they came to do."

Once lunar explorers finish their workday, it's time to go home. But what will home look like? How does one arrange living areas near dangerous landing zones? Where does the power come from? In short, how does one design the first settlement on another world?

SHACKLETON: HOME, SWEET HOME

Part of the answer lies within Building B220 of the Johnson Space Center. In the high-ceilinged warehouse-like structure, engineers are busy fabricating low-tech mockups of habitats that could comprise the international outpost at Shackleton. The newest mockups consist of Styrofoam and plywood, making modifications and design changes simple and inexpensive.

Several approaches to habitat design are under consideration. These fall into two general categories: rigid structures and inflatable ones. Lunar architects designing rigid habs have many decisions to make: should the structure be cylindrical? If it is, should it lie horizontally or stand vertically? The structure might also be configured as a "tuna can," using the full diameter of the Ares V to create large, open spaces in a low-lying structure.

Cylindrical arrangements are smaller and lighter, and thus easier to transport. Once a cylinder is on the ground, the interior arrangement is fairly inflexible, with bunks, plumbing, electrical components, and shelves attached to the structure. Multiple cylinders would be chained end to end, forming long corridors of work/living areas in linear arrangements. A tuna can, on the other hand, provides a circular floor that can be reconfigured into labs, sick bays, eating areas, etc. The main drawback of the tuna can is mass: it's a heavy payload. And while tuna cans might accommodate individual crew quarters (as opposed to the sleeping bunks of a cylindrical hab), engineers have a high degree of confidence and experience with the cylindrical model. They've been using them for years as building blocks of the International Space Station. "We're juggling a lot of tradeoffs," says lunar architect Robert Howard. "Both versions are built to be carried by *Altair*, but if the interior space is too small, you have psychological problems with the crew. If it's too large, the crew is fine, but missions are shorter because you can't bring as many consumables."

To hit the right balance, Howard is in the process of building several of the Styrofoam/wooden structures of different sizes. Although the modules of the International Space Station are 4.2 meters (yards) in diameter, they are scaled to fit the space shuttle cargo bay. The shuttle will be long retired when the habitats reach the lunar surface, so the new designs call for something a bit smaller and more mobile. Engineers are experimenting

Working mock-up of a tuna can habitat.

Exterior of a rigid habitat mockup. (photos by author)

with habs that are both 3 meters (yards) and 3.5 meters (yards) in diameter (roughly as far across as a Boeing 737 airliner). On paper, the two versions are so similar in scale as to suggest equality. But step inside the mockups, and the half-meter difference is remarkable. Taller ceilings give an impression of a larger floor area. It's an old trick: Architect Frank Lloyd Wright used this perceptual phenomenon to his advantage in small home designs: low-ceilinged hallways open into rooms with higher walls, giving the impression of a great expanse in a relatively modest living area. With an extra half meter of diameter, lunar inhabitants will be able to do simple things such as open both sets of under-bunk drawers at same time. But size and carrying capacity of the lunar lander are all in flux, says Howard. "Habitable volume and payload capability of the *Altair* are going to be fighting a battle for the next several years."

Plywood and Styrofoam interior mockups enable designers to assess living spaces without the use of expensive materials. Lightweight elements also allow for easy changes.

Many NASA experts would like to see habitats similar in design to the rigid modules on the International Space Station. They are heavy, but they work. Both NASA and Russian experience with long-term space habitation is within the microgravity of Earth orbit. Designing for a zero gravity environment, every surface can be used. There is no ceiling or floor, no up or down, so the physical arrangements are quite different from an environment with floor, walls, and ceiling. As one designer put it, "You're not just providing volume any more; it's a footprint."

Still, designers are weighing their options, and another one of those options is to scrap the idea of a rigid hab completely in favor of an inflatable one. An inflatable habitat has a rigid core containing supplies, electronics and other equipment, cocooned within a deflated habitat. Once on the lunar surface, the core would pressurize the donut-shaped habitat around it, with the core representing the donut "hole." Designers are experimenting with systems that can unfold floors and ceilings as the hab extends outward into a tuna-can shape. Layers on the exterior must include micrometeorite protection, thermal insulation, and a restraint structure to hold the pressurized "balloon" in a workable form. Another configuration of an inflatable hab would resemble the cylindrical shape of a rigid hab. Several industry partners are now at work integrating a solid hatch on the ends and side portion of inflatables.

Technicians at Houston's JSC are designing a pressure bladder inside a web of Vectran® straps (similar to Kevlar). The supporting Vectran webbing is then cut in several critical areas to test how the structure holds up under pressure when some of the outer structure fails. The advanced work has

A 2/3 scale inflatable habitat. Exterior Vectran® strips, used here to test pressure when some strips fail, would be covered with micrometeoroid and thermal blankets on the lunar surface. (photos by author)

already provided benefits to society, as industry gains insights into high-stress commercial materials and failure modes for manufacturing.

In the early developmental stages of the ISS, ILC Dover Corporation worked with NASA to design and fabricate Transhab, an inflatable habitat originally intended for use in interplanetary travel to Mars. NASA considered Transhab for use on the ISS, but Congress pulled the plug. House Resolution 1654 banned further development of inflatable habitat technology as it applied to the ISS. The official reason given was that due to cost overruns on ISS, transhab should not be developed alongside the rigid modules already in progress. Some analysts suggest that the cancellation was due to the reluctance on the part of some in government to allow technology that might promote interplanetary exploration (the bill's primary sponsor specifically referred to his desire that no funds be spent on human interplanetary capabilities). Whatever the causes, Transhab seemed dead on the ground until Bigelow Aerospace bought various rights and patents from ILC Dover to continue its development under a private industry flag. Bigelow currently has two 1/3 scale inflatable structures in orbit and plans to orbit more.[17]

Meanwhile, work still continues on inflatables at ILC Dover, the most visible element being the inflatable habitat research ongoing at McMurdo. ILC Dover has fabricated an inflatable habitat that covers an area 16 feet across by 24 feet in length. The entire system weighs about 1,000 pounds. In comparison, a rigid habitat tips the scales at well over 5 tons. In recent tests in Antarctica, four people assembled the structure in less than an hour. The undeployed package takes up a volume of 128 cubic feet, equivalent to the interior space of a Ford Expedition. When deployed, the structure yields

17. For new images from orbit, see http://www.bigelowaerospace.com

a living area of 2,500 cubic feet, an expansion ratio of 20:1. The habitat was deployed in January of 2008 and is outfitted with instrumentation for temperature, pressure, power usage, and gas leakage. The structure is being remotely monitored at Johnson Space Center and will be on line for a minimum of one year.

(l) ILC Dover's inflatable habitat in Antarctica; (r) spacious interior of the ILC hab. (images courtesy ILC Dover)

Johnson Space Center's Chris Culbert is weighing the rigid vs. inflatable habitat options. "They bring different strengths and weaknesses to the table. It really depends on the nature of the activities you're performing. Inflatables are nice for volume, so for human habitation space that works very well. But when you want a laboratory structure with well-defined stations for certain kinds of work, you may need the hard-shell to give you some of the equipment infrastructure you require. Ultimately, like Space Station, we'll have multiple modules, and it wouldn't surprise me if we had some that have inflatable portions and some that are rigid. We may well have different approaches by different nations."

NASA's lunar architect Larry Toups also envisions a mix of rigid and inflatable structures. He likens the variety of ongoing studies to movies on DVDs. "You know how you get these movies where you have the three special secret endings that were never used? This is like that. We have a story to tell here, and we don't know which ending or endings we'll finish with. One of the different endings: send a series of rigid habs. You can stay in one for a short while, but you need to add two more habs to create a good, workable living space so you can function with four crew members for 180 days."

Once a habitat is landed, it may have the capability of fending for itself for some time. Mission planners would like to build modules capable of self-deploying ahead of the crew. This will be very important for future exploration at Mars. Habitat autonomy would entail the use of ATHLETEs or similar remotely controlled rovers. These robot workhorses would off-load the hab from the *Altair* cargo lander, drive the hab to the outpost site, and set the module down. Larry Toups lays out the scenario: "You send a habitat ahead of the crew, have it on the surface, flip the switches that you need to remotely, and have it where it says 'Okay, we can have crew members move in here now.' So this ending has you send a large structure that is landed in one piece and can support a crew of 4 for 180 days." Toups calls this the Mars forward-looking approach. It's the strategy that a Mars mission would require.

The autonomous habitat concept is controversial. Some, like Marcia Ivins, are skeptical that assembling habitats robotically will be advantageous. "If you land an element on the surface, what seems to be a good idea is that you simply move that element on a flatbed truck or drop the wheels and drive it. To do it robotically, you have to have already placed on the surface the robotic system that you have to operate, so what would be the point? It's just an inefficient way to do it." Like so many other aspects of lunar outpost design, great minds will be pondering robotic vs. human-tended base deployment for years to come.

How many habitats will a permanent human presence call for? NASA's Larry Toups has been studying the problem for some time. The answer is complex. "The number of modules depends on not only the number of crew but also length of surface stay. With 4 members, you get a break at 30 days. Up until that, you are in camping mode. You hit 30 days up to 6 months, and you start having to provide additional volume, resources, food, clothing, consumable gases, and so forth." Even the smallest habitats have minimum requirements; they have bunks, but not dedicated crew quarters (in the case of cylindrical habs). Each habitat must house subsystems for life support and power management. The first modules might have small laboratory areas for sample analysis. But these evolve to an outpost by adding additional volume, enabling functions to be moved from the cramped initial area to dedicated areas for research, exercise, etc.

Some early modules might initially be packed with consumables. "The notion is that you deliver a core hab and logistics (supply) modules that plug into that," says Toups. "In essence, the crew eats their way into a habitable volume."

But for a skeleton crew of four to stay for 180 days, at least three habitats will be required for safety and health. What those habitats eventually consist of is anyone's guess. Toups suggests that, "Our initial footholds will probably use systems and technologies that we are comfortable with. Growth will come from that, evolving from construction shacks to more complex structures."

"It's important to understand that we don't have any kind of baseline for a lunar outpost yet," cautions Chris Culbert. "The work we have done so far is at the concept level only, and we have put together info on a lot of different kinds of concepts. There are themes that obviously emerge from the concept work that seem likely to occur in almost any kind of outpost scenario, such as a place for humans to live, a way for humans to get about, a way to generate power, etc. But the details vary considerably depending on what approaches you choose or which variable you want to optimize."

The outposts will provide the essentials of life, such as food, water, and air, but they must serve another important purpose: protection from radiation. Earth is enshrouded by strong magnetic fields generated by a core of molten rock and metal. The magnetic fields, known as the Van Allen radiation belts, funnel dangerous radiation away from the surface. Some of it ends up at the planet's poles, resulting in the ghostly glow of aurorae. The ISS and shuttle flights orbit well below Earth's shielding magnetic bubble. But the Moon has no such protective energy fields. Habitats must provide shelter from incoming radiation, and accomplishing this is not as easy as it might first appear.

Lunar explorers must concern themselves with two types of radiation. The first comes from the background radiation of cosmic rays. *Apollo* astronauts reported seeing flashes of light in the darkened cabin, a result of cosmic rays passing through the retina. Long-term exposure to cosmic rays can cause cancer or other health problems. Cosmic rays have such high energy that no artificial material significantly blocks it. One good thing, however, is that from the surface of the Moon, fully half of the cosmic rays are blocked by the Moon itself.

The second type of radiation occurs less frequently but is far more deadly. It is the high energy radiation that explodes from the Sun's surface as solar flares. Solar radiation can be filtered out with something that is readily available on the Moon: dirt. Several meters of regolith may be needed (tests are still underway), but a habitat can be designed to carry the load on its roof. Even an inflatable enclosure can be built with regolith protection in mind. NASA's Langley research center is developing a Quonset-hut style structure that starts out flat on the lunar surface. Shackleton construction

A Chariot rover equipped with a blade for moving lunar regolith. Note the second gondola control station mounted on the far end. (photo courtesy NASA)

workers would scoop dirt or load sandbags onto the flat covering, then inflate it into an arch. A habitat would then be deployed underneath. "We've seen some data that regolith is effective from a radiation standpoint," says JSC's lunar architecture guru Chris Culbert. "The flip side of that is that moving around many tons of regolith on the lunar surface is non-trivial. We'd have to take earth-moving gear with us." Teams at JSC and Kennedy Space Center are studying a bull-dozing blade that can affix to various rovers under development.

A 2-meter layer of regolith affords insufficient protection from the most dangerous of events. Some types of metals or a combination of metal walls and water storage may do the trick, as would an underground storm-shelter. Studies have been under way for several decades, with options still wide open.

Water makes an efficient barrier to radiation and can be stored in containers within the walls of living areas. Some types of plastics are more radiation "hardened" than others. Analysts are still shopping.

FIVE BARS ON YOUR PHONE

Like McMurdo base before it, Shackleton will eventually develop into a multi-structure, sprawling community with roadways, centers for special-ized operations, and staging areas for expeditions to the ends of the Moon. But just chatting with someone over the next hill can be a challenge in the Moon's airless environment. Radio communication can be hindered or cut off by rocky embankments or curves in the road. Communications within the lunar community promise to offer some of the greatest challenges to daily living and working on the Moon. At a minimum, lunar settlers need the ability to communicate with Earth, and to communicate with all the habitats, astronauts, and rovers within their local community. Inhabitants will need both voice and video. High bandwidth capability is preferable for downloading scientific and engineering information. Repeater stations or communications towers might fulfill Shackleton's requirements. All ele-ments of the local outpost itself will remain within line-of-site, but scientific and exploratory interests will quickly take inhabitants over the hill. Sortie missions need to be able to communicate from wherever they land, and sur-face operations require links to orbital or incoming *Altair* crews. Traveling rover crews might deploy a series of towers—along with supply caches—as they roam across the lunar wilderness.

Strategists continue to weigh options for surface vs. orbital systems. If a single communications satellite is placed in polar orbit, lunar settlers could have coverage up to 40 percent of the time. A second satellite brings coverage up to 80 percent. An entire constellation of four or five increases capability dramati-cally, offering not only services such as continuous communication but also tracking and data, something like a lunar GPS system. International partners

Being There

What is it like to hike across a lunar valley, or scale the mountains of the Moon? The lunar environment is immediately recognizable as alien, in aspects ranging from obvious to subtle. Astronauts describe rocks bejeweled with crystals. When the Sun is behind the observer, explorers tell us, the regolith sparkles like snow. Boulders are strewn, as if by a giant's dice-toss, across hills and plains, resting since ancient impacts pitched them into the Moon's Sun-drenched ebony sky. Stones everywhere are covered in tiny white dots, blemishes from micrometeorites. The landscape is pocked with craters of every size, down to small rimless bowls, each with a fused-glass center. Here are some other first-hand observations:

"You've got craters that all look the same. When you're standing next to them, you're saying to yourself, 'Is this the crater that's 300 feet wide on my map, or this one that's 200 feet, or the one over here that's 400 feet?' You never really know. The next thing is your movement. The 1/6 g is very different. You're not familiar with how far you can run or walk in a given time. You couldn't tell if you'd run half a mile or a quarter mile. You didn't have the same reference."

Alan Bean, *Apollo 12*

"Unlike earlier flights, we were in the highlands. We were in a plateau area, not marked by mountains like the later missions. It was very much like sand dunes with craters. That's pretty spectacular in itself. The regolith is the consistency of talcum powder. It's packed, so you don't sink into it far, but it's so fine-grained. It's incredible how soft it is. The edge of craters were softer. It was rough enough terrain that we had difficulty in finding our targets. Seeing the patterns in overhead photographs simply weren't available to us on the surface. It's a lot like being in the desert, where distant objects seem closer. That's magnified on the Moon because it's smaller and distances are distorted. You could almost see the curvature of the Moon because of the smallness of the planet itself, unlike Earth, which is four times as big. It was very confusing trying to get your bearings."

Ed Mitchell, *Apollo 14*

When we got up there and turned around, I thought, "Man, we are really on a steep hill." It felt like we'd fall out the front of the rover. The view from up there was breathtakingly beautiful. You could look all the way across the Cayley Plain to Stone Mountain to the north of us. To the northwest and west was an endless view of rolling terrain, all the way to the lunar horizon, which was very sharp since there was no atmosphere. It was generally shades of gray, with the South Ray Crater being real white. The only other color was the little lunar module *Orion* sitting in the middle of Cayley Plain, reflecting its yellow Mylar. The gray and black of the lunar module was spectacular.

Charlie Duke, *Apollo 16*

"It was a combination of all aspects of the valley being deeper than the Grand Canyon. The mountains on either side rose to 6,000 and 7,000 feet above us, brilliantly illuminated by a Sun as bright as any high Colorado mountain Sun you can imagine, but with a very sharp contrast in that the sky was black. These mountains were outlined against the black sky rather than a blue one. I think that all of that, with Earth over the southwestern mountains, all created a single image that I'll never forget."

Harrison Schmitt, *Apollo 17*

such as the European Space Agency and India have developed complex, advanced communications satellite constellations, and they are a likely source of such a global lunar communications network.

Whatever communications system is finally used, it is likely that the system will also serve in a navigation role. Without Earthly cues of scale and landmarks, just finding one's way around is a challenge. For example, on Earth, the atmosphere tends to shift the color of objects as they increase in distance from the observer. This phenomenon, called atmospheric perspective, is a key element used by human perception to judge distance. But the lunar environment is airless, explains *Apollo 12*'s Al Bean. "Because there's no air, you can't judge distance or size." *Apollo 14*'s Ed Mitchell agrees. "The only major problem we had was finding landmarks by eye. That turns out to be very difficult to do."

Lunar inhabitants will need to rely on both visual perception and technology to negotiate the rugged highlands of the South Pole. Shackleton inhabitants may equip their rovers with a system similar to a TACAN device, where the vehicle continually keeps track of its movements in comparison to its point of origin. Communications beacons can also be used as local positioning devices, networking radio signals among vehicles, outpost habs, communications stations, and other waypoints.

Like so many other elements of a lunar outpost, communications take power. Shackleton base may link its infrastructure to a field of solar energy collectors. These may be conventional photovoltaic cells or solar heat exchangers of some kind. Lunar dark periods require power storage, so battery technology is yet another area of research under pursuit by lunar architects. Some portions of an advanced outpost might require low-level nuclear power sources, and advanced rovers designed to brave two-week lunar nights will surely need power sources other than solar ones. Chris Culbert believes power can be managed with the long daylight periods at a site such as Shackleton. "In lighting conditions at the poles, you don't need to store energy for more than hours or days at a time. You can also play operations games where you define when to send people. For example, we can say, 'Okay, during these next six months, we only have three periods of darkness, and the longest one is only fourteen hours. That's fine; we'll send humans during that six-month period. But in the five months after that, there's going to be fourteen periods of darkness and one of them lasts four and a half days, so we just don't have the energy storage in place to do that with humans there yet.'" Initially, any outpost elements will be self-sufficient, containing enough energy storage to keep the systems alive when nobody is there. With human presence, systems will use much more energy. The timing of missions can maximize power usage for the systems already deployed. Culbert says, "Over time, with our resources and those brought by our international partners, you want to build up infrastructure so that humans can stay permanently."

The series of habitats, power generators, construction shacks, and rover enclosures will need to be situated a safe distance away from incoming payloads. Planners envision a landing area hundreds of meters to several kilometers from the inhabited base. *Altair* cargo and crew carriers would touch down in the landing area,

Sorties in *Apollo* and Constellation

The lunar roving vehicles revolutionized *Apollo* exploration, making possible "sorties" away from the home base of the lunar module. But while the *Apollo*-era rovers expanded exploration to kilometers, NASA's plans for sorties in the Constellation Architecture will encompass excursions of hundreds or even thousands of kilometers. New NASA plans call for the capacity to send four astronauts to any site on the Moon for short periods, giving crews full global access to both the near and far hemispheres. Sorties would last up to seven days, with EVAs possible every day. Unlike *Apollo*'s lunar module, the advanced *Altair* will house an airlock so that the main cabin remains pressurized throughout the mission. Crews will exit through the airlock to access the surface and can explore in two teams of two, thus increasing efficiency. With a crew of four, *Altair* will have capacity to carry 2 metric tons, a tenfold increase over *Apollo*'s LM. Taking the place of *Apollo*'s open-cockpit lunar roving vehicles will be advanced long-range rovers, some of which may be pressurized for multi-day journeys.

perhaps guided by landing beacons or—in the case of unmanned craft—remotely piloted from outpost residents. Their contents would then be ferried to the crater rim outpost aboard Chariot- or ATHLETE-style rovers. Roads might be graded between the sites, but packing dirt is impractical in a vacuum. A more plausible solution would be to line driving routes with thin grates that partially elevate traffic off the dusty lunar surface. These grates might be only a few centimeters deep but would be enough to keep dust to a minimum.

ANATOMY OF A SURFACE EXPEDITION

Even after an outpost is established at the lunar pole, sortie missions will continue, perhaps funded fully by universities, national organizations such as the British Museum or the National Science Foundation, or international groups such as the European Space Agency. Some missions will take the form of *Altair* sortie flights, where a lander touches down in a remote area with a small, *Apollo*-class rover for a week. Other expeditions will be outfitted at the outpost for longer journeys. These will undoubtedly make use of multiple vehicles for safety and flexibility. Large rovers will be delivered on *Altair* cargo shipments. These pressurized vehicles will serve as roving habitats and may have ranges on a scale of thousands of miles. They will carry—or more likely pull—smaller vehicles to explore rugged terrain away from base camps. JPL's

The "Winnebago approach" to exploration. Here, the large vehicle serves as primary transport and living area—the habitat—while the smaller vehicle is reserved for shorter, more specialized exploration.

Brian Wilcox calls this strategy the "Winnebago mode." "When you see a Winnebago towing a Jeep down the interstate, you immediately know what they plan to do; they will get to a good camping spot, set up the Winnebago as a habitat, use the Jeep for day-trips radially out from the mobile habitat until that locale is explored, and then pack up the whole assembly and move to a new camping site."

IT TAKES A VILLAGE

If all goes according to plan, a vibrant outpost of habitats, roving vehicles, communications systems, and power infrastructures could be in place two decades after the first humans return to the Moon's surface. It promises to be a village that would make Ernst Shackleton proud. Chris Culbert envisions a robust international community. "You can use the Antarctic as a well-defined international model for getting people and assets there. Commercial entities are taking advantage of the infrastructure already in place; maybe commercial entities are running it. You might get stationed there for six months to finish your dissertation, for example. Just as in Antarctica, there will be people who set you up, outfit you with gear, and train you on how to do business there. Eventually you get very healthy private enterprise, perhaps setting up a hotel so you can go stay there for a week. In the meantime, NASA may have set up a lot of the original infrastructure, but they've moved on to do other things, perhaps setting up outposts on Mars or the moons of Jupiter. Who knows? But twenty years after we've begun, all this strikes me as very possible."

In addition to advanced technology, an advantage that new explorers will share with *Apollo* astronauts is experience. Regular missions will build on each other, so crews can hone techniques as they do lunar construction and exploration. *Apollo* astronauts demonstrated a sharp learning curve, culminating in the efficient confidence, even boldness, of *Apollo 17*, the last flight. "Our confidence came from knowing that others had done so well, and that the equipment and procedures and the operations were really finely tuned by the time we went," says *Apollo 17* LM Pilot Harrison Schmitt. "That applies to the entire team of flight controllers and scientists and everyone else. We were working with a great deal of efficiency compared to the earlier missions. We just were better prepared. We were better trained; we knew what to expect."

For the new arrivals at Shackleton, it will not be business as usual. The first time a person faces the alien environment of the Moon, no matter how well they have trained, will be a peak life experience, says veteran shuttle astronaut Marcia Ivins. "If you talk to enough EVA guys, the very first time they go out they get that same feeling: this is not like being in the [simulations]. There is no way to prepare for it or train for that. You just have to recognize that it's going to happen to you, that it will be a moment, it will be your moment, and there will only be one like it. Savor it and enjoy it for what it is."

But the glorious vision of glistening Moon modules and polished rovers is not enough. The entire Shackleton outpost scenario begs the questions: Why should we go back? And aside from the technological, social, and political reasons, there are scientific ones. Lunar science will drive our knowledge base while providing practical lessons for society.

The Moon may have been birthed by a catastrophic impact between Earth and a Mars-sized planetoid.

Chapter Five

Scientific Reasons to Return

They were there for all to see, but no one could read their secrets. Scattered across monuments four millennia old, cut into stony hillsides, and painted on the walls of the pharaonic tombs, the language of the ancients teased archaeologists, tempted them to false conclusions, frustrated them at every turn. What were the Egyptians trying to say? In the summer of 1799, a French expedition uncovered a slab of basalt 1.1 meters high and broke open the entire field of Egyptology. The Rosetta stone was inscribed in three tongues: the mysterious Egyptian heiroglyphs, a form of simplified Egyptian writing called demotic, and ancient Greek.

French linguist Francois Champollion, building on the earlier work of Thomas Young, concluded that the three-language tablet constituted three parallel decrees by Ptolemy V. Champollion realized that the hieroglyphs were not only representational picture writing but also phonetic, using combinations of sounds to express words. Champollion's breakthrough-discovery marked the very beginning of modern Egyptology.[18]

Earth's Moon is the Rosetta stone of the terrestrial planets. Across its basalt face, the Solar System has inscribed its history in languages that we are only beginning to understand. Translating those languages will open the door of time, answering many questions. Questions abound. Where did the Moon come from? What is it like beneath its battered surface? Can Earth's natural satellite fill in the blanks of the earliest history of our own world and its inner planet siblings? Thanks to the Moon's rock-hewn story, humanity will gaze back to the beginning, when molten planets weathered a hail of rock and iron from the Solar System's formative years. The lunar surface displays this history more clearly than any other site in the Solar System, says NASA Goddard's Chief Scientist, James Garvin. "The only place where those events are recorded and not modified much by anything other than space weathering (the action of cosmic background radiation, solar wind, and micrometeorites) is the Moon."

Like all planetary science, understanding the Moon will better equip us to understand and care for our own planet. But lives may be at stake in the nearer term. To safely establish a permanent encampment on the Moon, we must understand our cosmic neighbor in greater detail.

A LITTLE LUNAR HISTORY

Thanks to discoveries from *Apollo* and the Soviet *Luna* explorers, lunar researchers have a good general view of the Moon's development. The entire Solar System coalesced from a disk-shaped cloud of dust and gas, called an accretion disk. The densest part of the cloud, at the center, collapsed in upon itself from the weight of its own gravity. Immense pressures and rising temperatures triggered a chain reaction that ignited nuclear fusion, the energy that powers the Sun. Eddies in the outer cloud contributed to rubble piles already forming, and planets began to appear. Mountain-sized rocks—and

18. Champollion published his work in the historic *Letter a M. Dacier…relative a l'alphabet des hieroglyphs phonetiques employes par les egyptiens* ("Letter to M. Dacier…on the alphabet of the phonetic hieroglyphs employed by the Egyptians") in 1822.

M. Carroll, *The Seventh Landing*, DOI 10.1007/978-0-387-93881-3_5,
© Springer Science+Business Media, LLC 2009

The dark maria regions of the Moon are remnants of vast lava flows. (Galileo photo courtesy of NASA/JPL.)

even Moon-sized ones—regularly slammed into each other as the Solar System sorted itself out. Large bodies gained mass and gravity, pulling in ever more cosmic debris in a runaway gravitational growth spurt.

Within 100 million years of its birth, the Sun went through an energetic epoch called the T-tauri phase. Solar wind shoved light volatiles out into the distant regions of the planetary system, leaving only the heavy materials behind.[19] Those materials formed the building blocks of the rocky inner worlds. As those worlds settled into spheres under their own gravitation, denser material settled to the core, while lighter material rose to form light crust in a process called differentiation.

As planets and large moons cooled, mineral-rich molten rock escaped to the surface, spreading out as lava flows. The dark blotches blanketing our Moon are remnants of magma seas that once simmered across its face. The bright rock from the ancient, cratered highlands represents the primitive crust of the Moon that solidified nearly 4.5 billion years ago. This crust consisted of solid rocks resting atop a deep layer of melted magma. For 600 million years, a hail of cosmic debris—meteors and asteroids—pummeled the crust into a cratered landscape that includes the highlands we see today.

Their surfaces scarred by stony asteroids and icy comets, some inner worlds began to resurface themselves in a variety of ways. The crust of Venus exhibits folds and wrinkles from internal forces called tectonics. Cliffs and faults score the faces of Mars and Mercury. All of the inner planets and Earth's Moon appear to have gone through at least early volcanism and magma oceans. Venus and Mars built extensive volcanic structures over long periods of time. All of these forces have visited themselves upon the landscapes of Earth.

19. Recent work by researchers such as Hal Levison at the Southwest Research Institute indicate that mixing from the inner to the outer Solar System took place during this time, bringing some icy outer materials back to the inner Solar System. As with most science, reality is more subtle and varied than models sometimes suggest.

Volcanism resurfaced the primordial Moon. A looming Earth—closer at that time than it is today—had no familiar continents. Instead, landforms were raised rims of impact basins and craters. (painting by the author)

In many locations, asteroids and comets punched through the crust, freeing magma to flow on the surface as lava. The larger craters and immense impact basins were later filled by lava flows, some spreading more than 100 km in length. Extensive lava flows probably began just after the heaviest rain of asteroids, about 3.9 billion years ago. Radiometric[20] dating shows that most flows occurred from 4 billion to 3 billion years ago. Lavas of great mineral variety resurfaced vast areas of the Moon, mostly on the near side. Volcanism gradually declined, probably ending about 1 billion years ago. The surface that is left today is a nearly unchanged version of the lunar world over a billion years past.

THE SKY WAS FALLING

To NASA's Garvin and others, the Moon is the ultimate control experiment. In the forty years that spacecraft have been exploring the Solar System, researchers have come to understand several common themes that govern the solid worlds. Scientists see patterns and commonality in the assembly, the destruction, and the modification of all solid planetary surfaces. While only some of the planets have had varying degrees of volcanic activity, weather, or plate tectonics that modified their surfaces, all of them have been influenced by hypervelocity collisions. Objects left over from the birth-cloud of the early Solar System or blasted from colliding objects were swept up by the gravity of nearby planets and moons. These small bodies traveled at tremendous speeds, and their impacts sculpted the planets' shells. "The process is so ubiquitous

20. Radiometric dating compares the amounts of unstable radioactive elements as they change to stable ones. Scientists are able to chart dates using these ratios. The most commonly known radiometric technique is carbon dating.

The KT boundary is a worldwide layer of iridium that atop the fossil sequence, including the dinosaurs. Below this layer are the dinosaurs and other life forms of the Mesozoic Era. Above it, there are no such life forms. Iridium is a metal rare on Earth but common in asteroids. The KT boundary first gave scientists a clue that periodic impacts may lead to the extinction of many life forms on Earth.

through the Solar System that we really need to understand it," Garvin says. "To me, understanding the role of collisions in planetary formation, evolution, evolution of life—if there is any—and ultimately the way planets grow up and evolve is tied to impact. The entire collisional history of the Moon, from the so-called cataclysm[21], through the mega-basin forming impacts such as south pole Aitken, Nectaris, Imbrium, Serenetatis, and onto Orientale, to the last major impacts recorded (by Tycho, Copernicus, Aristarchus, etc.) is special. It shows how a planet's collisional history can affect the evolution of its crust, the role of volatiles (such as water), and ultimately how major impact cratering works on any planet. Most of the record of this stuff has been eradicated on Earth, buried on Venus, and modified on Mars. So the Moon is key."

Wendell Mendell, Chief Scientist of NASA's Constellation program, agrees. "We've studied the Moon and we understand where it sits in the lineup, and if we go probe the historical development of the Moon and the processes of the Moon, then we can get some insight into the other planets and Earth, because on Earth the early history is gone; on the Moon it's preserved. Whatever the Moon has seen in terms of external environment in the past, Earth has seen, but on Earth it's erased. It's only in the last twenty years that we've come to appreciate the fact that we live in an environment where things go plunk, and that there might be a reason that 99 percent of all species that have ever lived on Earth have gone extinct." If space rocks that "go plunk" have affected past life, might they not also affect modern life?

The influence of these impacts is unmistakable, not only to planetary scientists but to those who study other sciences. Paleontologists see plentiful evidence that impacts in the past have resulted in mass extinctions here on Earth. David Kring, lunar geologist at the Lunar and Planetary Institute, says, "We want to use the Moon as a clock, to help us see what effect impacts had on life on Earth. After all, the KT (Cretaceous/Tertiary) impact[22] had an effect on life; it wiped out the dinosaurs." The connection between impacts and

21. An epoch of dense meteor bombardment that appears to have tailed out roughly 3.8 billion years ago, about a billion years after the Solar System's birth.

22. The KT boundary is the layer of sediment representing the end of the age of dinosaurs, called the Cretaceous period. It is high in iridium, a substance common in meteorites.

our own well-being is of paramount importance. In a very real sense, the Moon has raised humanity's cosmic awareness. *Apollo 8*'s historic views provided an awareness of humankind's place in the universe, and the scars of lunar impacts have given us insights into the dangerous nature of the space environment around our delicate world.

THE MOON IN ITS PLACE

"We now understand that the Moon is a differentiated body," Mendell says. "It's kind of a planet, it's arrested in its development, and it can be seen as an end member of the terrestrial planet evolutionary scheme. Within the terrestrial planet system, you can look at these things of different sizes as illustrating different stages at which the process stops. Earth is the most evolved of all the planets."

Unlike the Moon or Mercury, Earth is still geologically active. It has internal heating through radioactive decay of material in its interior, along with some heat left over from its formative, hot years. It has a magnetic field, thought to be generated by a still-molten iron core. Only Mercury shares this feature, but it is not clear whether Mercury's magnetic fields are generated by an internal dynamo (like Earth's) or is merely a record of earlier internal magnetism frozen into the rocks as a remnant of a long-dead core. This geologic fossilization of magnetic fields can also be seen in the rocks on Earth's Atlantic Ocean floor. There, at the planetary seam where new rock spreads away in opposite directions, geologists see mirror images of magnetic patterns frozen into the rocks. The orbiting Mars Global Surveyor has detected hints that Mars may have such a mirror-image record of ancient magnetism in its rocks, but studies are still under way. Even if so, Mars' magnetic core appears to have cooled and frozen long ago.

Earth also has an active atmosphere, as does Venus, Saturn's moon Titan, and, to a lesser extent, Mars. Weathering shapes the surfaces of these worlds, chiseling canyons, eroding mountains and crater rims, and grinding out vast sand dune fields.

Finally, Earth has a biosphere. It appears to be alone in this feature. Certainly, no other terrestrial planet has been influenced so heavily by biological activity. "So you have a series of check-boxes," Mendell explains, "and all the boxes are checked on Earth, but the others have varying amounts of checks. The question is why? Some differences can be explained by varying distances from the Sun, but that's not the whole story. Some of it has to do with size. Interestingly enough, some of it has to do with the fact that Earth has a large satellite, which none of the others do. The Moon tends to stabilize the polar wander, which Mars suffers from."

The Moon also creates tides, which many biologists feel led to life. With Earth on one side of the planetary evolution scale, and the Moon on the other, the Moon may offer the most promising laboratory in which to discover the true nature and history of the inner Solar System. As Jim Garvin

Doing science, Apollo *style: Each* Apollo *expedition carried an* Apollo *lunar surface experiments package (ALSEP).* Apollo 17 *carried the following instruments (left to right): heat flow experiment; deep core neutron flux; (lunar rover, lander in distance), RTG (foreground, nuclear power source), and at far right, the ALSEP array, including the LEAM (lunar ejecta and meteorites), the LSG (lunar surface gravimeter), the large central tower with antenna, and a network of low-lying geophones to sense Moonquakes. (Photo courtesy of LPI/NASA.)*

puts it, "You might be able to find this stuff out on Mars or Venus, but they are much more complicated. You've got big atmospheres, the role of water, the role of massive resurfacing in the case of Venus (we think), and history of oceans. All these things that complicate the issue on Mars and Venus we don't even have to worry about on the Moon. For me, the Moon is a laboratory for that and many other things. It's as if Mother Nature said, 'Hey, come learn about this by going to your nearest place.'"

UNSOLVED MYSTERIES

Despite the successes of *Apollo* and the *Luna* programs, many fundamental questions remain to be answered. Here's a sampling:

What is the Moon's crust like beneath the surface? Orbital data only shows the top layer, and craters only offer limited windows into the interior. To truly understand the deeper layers and their history, deep samples must be obtained, either by core drilling or by exploring large impact basins that may be littered with rocks from deep inside.

Does the makeup of the crust vary from the near to far side? It is obvious that the crust on the near side of the Moon is thinner than that on the far side. The near side is blanketed by flat maria, plainlike areas where impacts broke through the crust to allow flooding of magma from within. On the far side, there are few maria. When scientists chart the path of satellites in orbit around the Moon, they can create a gravity map that shows that the crust on the far side is, indeed, thicker. Why does the Moon have this dichotomy?

Do the sheltered craters at the north and south poles harbor primordial ice deposits? The floors of some craters may have been in shadow since ancient epochs. Their temperatures, hovering at a constant −233° C, are cold enough to trap water molecules from incoming comets and meteors. Water may be preserved at these sites, making them important resources for lunar outposts.

What can lunar craters reveal about the formation and history of the early Solar System? We have seen the relationship between the Moon and other terrestrial planets. Earth's nearest neighbor has much to tell us about the evolution of planetary surfaces, including Earth's.

Was the volcanism on the far side similar to that on the near side? Lunar geologists are just beginning to chart and recognize many volcanic structures on the Moon. Did the Moon's closer hemisphere play host to different levels of volcanism from that of the farther one?

Can the Moon fill in the missing history of Earth's earliest eras? Earth's most ancient rocks are about as old as the youngest rocks found on the Moon. A record of Earth's earliest years may well be written in stone on the Moon.

One of the great unsolved planetary mysteries surrounds the origin of the Moon. Simply put, scientists do not know how it got here. Many theories have come and gone as science accrues more data on the Moon's makeup and as computers become more powerful. But deciphering the languages written in the Moon's stone is as daunting a task as decoding Rosetta. It is as if scientists have disjointed pieces of a broken vase, with tantalizing hints as to how they go together. But the patterns don't quite fit and have led to many alternative concepts.

Early astronomers suggested that the Moon formed out of the same primordial cloud that Earth did. This "accretionary theory" is supported by another planet: Jupiter. Jupiter's four largest moons almost certainly condensed out of the cloud of debris that also gave birth to Jupiter. "It happened in some weird sense in the Jovian and Saturnian systems," James Garvin observes. "Why Earth got a big [moon] and Venus didn't we can debate. Some people think Venus had one and it collided with the parent, producing the strange motion of Venus[23], its loss of a magnetic field, and other aspects that are hard to explain." But there are problems with this theory as it applies to our own Moon, not the least of which is that it fails to explain why the Moon lacks iron. This puzzling fact came to light as *Apollo* and *Luna* samples came under scrutiny in laboratories across the world.

Another early concept saw the Moon as a wanderer, formed somewhere else in the Solar System where there was little iron. According to this theory, the Moon was later captured into orbit around Earth. Some theorists believe lunar rocks disprove this idea. When researchers finally had Moon stones in hand, some isotopes in the alien material matched isotopes within Earth's rocks.

A third popular scheme suggested that the early Earth spun so fast that the Moon was torn away from it, possibly from the Pacific Ocean basin. Although this model would explain the similarity of Moon samples to Earth's mantle, mathematics showed that the total angular momentum and energy to form the Moon in this way would not result in the motions of the Moon and Earth that we see today. Something was missing from all the theories.

Some astronomers believe they have come close to the answer. In 1975, William K. Hartmann and Donald Davis proposed a new scenario for the mysterious birth of the Moon. They proposed that an Earth-shaking event took place shortly after our planet became differentiated. Differentiation is a process that takes place early in planetary formation, when heavier material settles inward, while lighter material rises to become the outer regions of mantle and crust.

If proponents of the theory are correct, it was a cataclysm that nearly put an end to the world we inhabit today. The event transpired during a violent epoch some 4.2 billion years ago, early in the formation of our Solar System. In the midst of this asteroid demolition derby, Earth's mass became large

23. Venus rotates in the opposite direction to most bodies in the Solar System. Its day lasts a leisurely 243 Earth days, longer than its 225-day year.

enough—and radioactive materials abundant enough—to heat the core. But before Earth could settle down into a respectable planet, a Mars-sized behemoth came careening out of the darkness. If the angle of impact had been slightly steeper, Earth would have shattered like a dropped china cup. Instead, the stray planet hit a glancing blow. The titanic impact peeled away the lighter material from Earth's crust. For a brief time in geologic history, the again-molten Earth had a ring to rival even Saturn's. Within less than a million years, however, that ring of debris had become the Moon. Perhaps.

This model has several advantages over all others. First, it explains why the Moon has the same oxygen isotope ratios as Earth, while rocks from other parts of the Solar System differ significantly. This shows that the Moon formed in the same cosmic neighborhood as Earth. The theory also explains the Moon's low density and lack of iron. Earth's iron had already migrated to its core by the time the giant impactor tore off the planet's outer layers. Finally, it explains why the Moon's orbit is not inclined at the same angle as Earth's. If the Earth/Moon system had formed from the same cloud, the Moon would be orbiting in the same plane as Earth's equator. It is inclined more than 5°.[24]

Although he is a fan of the impact theory, Garvin cautions that the jury is still out. "There are certainly issues that have not been addressed. We shouldn't assume we know. We have one working theory that fits some data that's pretty cool, and that the dynamics people can make work. I suspect it is the easiest one to make work, but that doesn't mean it's the right one. We have more work to do to understand how large planetary satellites come to be. We have Titan at Saturn, we have our Moon, and the whole Galilean system at Jupiter[25], and we have working theories for these that a lot of smart people have come up with."

Underscoring the healthy controversy generated by the subject of lunar origins, *Apollo 17*'s astronaut-geologist, Harrison Schmitt, outlined reasons that he doubts the impact origin theory. He cites the presence of certain isotopes in lunar samples. "The main problems for the giant impact [theory] are the high volatile content of the…orange and green pyroclastic[26] glasses from *Apollo 17* and *15*, respectively. These should have been erased if that hypothesis is correct due to the extreme temperatures resulting from such an event." Schmitt points out that models indicate the Moon would have become molten to the core from the heat of a glancing blow. The problem, he says, is that there is good evidence—found in the isotopes nestled within some lunar samples—that the Moon was only molten down to a depth of 500 km. "That type of origin [from a giant impact] is recognized to create extremely high temperatures that would almost certainly destroy any volatile components such as we find in those pyroclastic glasses. The whole impact hypothesis comes from computer modeling. They're trying to make those computer models explain the geology. I sort of go the other direction. I say the geology has to be in the driver's seat, not the computer."

To researchers like lunar geologist David Kring, the big questions remain. "When *Apollo* flew, we were so ignorant that we only had a few questions.

24. This much of a tilt in the Moon's orbit suggests that the Moon's early path was inclined by over 10 degrees. Its interaction with the gravity of the debris disk itself may have contributed to its current inclination (see Canup and Ward, *Southwest Research Institute News,* Feb. 15, 2000).

25. The Galilean satellites are Io, Europa, Ganymede, and Callisto, named after their discoverer, Galileo Galilei.

26. Glasses formed in volcanically-related processes

Glasses like these, found in Apollo *lunar samples, indicate to some scientists that the Moon formed under cooler conditions than would result from a giant impact. (NASA/LPI photo)*

At the beginning of the *Apollo* era, scientists were still debating whether craters were of impact or volcanic origin. Despite a series of robotic landings, the thickness of the [Moon dust] was unknown. Would it hold up a heavy *Apollo* lunar lander? Would dust be a problem for landing or navigation on the surface? But literally thousands of scientific questions remain, and many were given birth by what we learned from *Apollo.*"

LOOSE ENDS

Mendell believes a return to the Moon promises to unlock many of these secrets, providing a quantum leap in knowledge. "You have to think about the jump from pre-*Apollo* to post-*Apollo*. We learned a lot, and there was a sense that now we could finally write the textbook. There was a perception even within the technical and scientific community that the problems were solved." But those optimistic declarations may have been premature. Armed with more powerful computers, the next generation of researchers is reexamining the decades-old *Apollo* data. They have discovered interpretation errors in the original analyses.

Seismic data is one case in point. *Apollo* astronauts left instruments called seismometers on the Moon to listen for "Moonquakes." The kind of information gleaned by these instruments helps geophysicists determine details about the structure beneath the ground, often as deep as the core of a planet. Seismic data gave researchers a ballpark estimate on the Moon's internal structure. But it is now generally agreed that the textbooks are wrong on the thickness of the crust and the discontinuities inside the Moon. The Moon's crust may be only half as thick as the textbooks say, which has profound implications for the interior structure and composition of the Moon, as well

 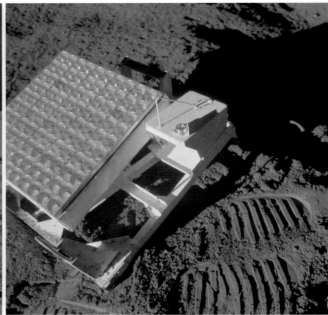

*Lunar instruments:
(left) A passive seismometer
left by Apollo 14 to search
for Moonquakes. The LM is
in the background, partially
obscured by a hill; (right)
Laser reflectors left by Apollo
astronauts are still in use
today, providing detailed
information about the Moon's
distance from Earth and
movement through space.
(NASA/JSC images)*

as the nature of its core. Mendell finds it interesting that researchers tend to look at seismic data to determine if a body has a molten core, as this is not the telling clue for the Moon. "The one piece of evidence that seems to say that there is something at least squishy in the center is not from the seismic data but from the laser retroreflectors that were left on the surface by *Apollo*, which are still being used today. There is so much data now, and the precision of the data is so great, that they can actually look at the motion of the Moon in space and determine the fact that it's not homogenous. It has structure and inside there is something a little squishy. It's almost magical in terms of the geophysics."

SHOCKING REVELATIONS?

Researchers are also still unsure about the nature of the lunar energy fields. The lunar samples all have magnetic properties, but the source of the magnetism is mystifying. Perhaps the Moon's rocks contain the same kind of magnetic "fossilization" that Mercury seems to. Other questions about lunar fields are more worrisome for would-be lunar inhabitants. One issue that has many designers concerned is the fact that the lunar environment is electrically charged. The Moon's surface is covered in what geologists call regolith. Regolith is pulverized rock, and differs from soils on Earth in that it contains no organic material. The lunar regolith builds up electrical charges tied to the day/night cycle. The Moon has no atmosphere and no magnetic field, so the surface is exposed directly to the solar wind. There are electrical fields in the solar wind, and data from spacecraft, particularly the *Lunar Prospector,* shows that on the daylight side, the net charge on the surface is positive, while on the night side the net charge is negative. Somehow, in between, the charge reverses. Moon planners see the charge difference from light to shadow and ask: Can an astronaut or equipment be damaged in the transition? *Apollo* astronauts explored equatorial regions and were in the uninterrupted sunlight for a short period (the Moon's day lasts about fifteen Earth days, so *Apollo* explorers spent a small fraction of a lunar day on the surface). But designers are considering permanent outposts in the polar regions, where astronauts will inhabit the surface during entire day/night cycles.

Apollo did not specifically look into this phenomenon. Scientists were more interested in the Moon as a small planet. Data was taken, but researchers are only now revisiting it to tease out the details, trying to pin down the structure of the lunar surface along with its electrical properties. Researchers have

The Moon is negatively charged in shadow, but positively charged in Sunlight. Will this pose a danger to future explorers? (Photo courtesy of NASA.)

only general answers. Mendell poses the important question: "If there's a big mountain that casts a shadow, is there a positive charge over here in the sunlight but a negative charge in the shadow, or is it only a planetary-scale thing? People are a little nervous, because they want more certainty than we can provide. The engineers don't like the answers we give them because we haven't taken a measurement at some certain spot. We can only talk about generic processes. We've only really been on the maria, and a little bit in the highlands, and they're talking about landing on the edges of huge craters at the south pole. The environments suddenly have uncertainty to them of the type you don't expect from the textbooks, and the engineers just want to know which encyclopedia to look it up in." Astronaut/geologist Harrison Schmitt agrees that much work still needs to be done, but he is skeptical that a daily passing of electromagnetic fields will be a serious problem. "I really don't think that we're going to find that charged dust particles move around very much. If that's happening with every terminator passage, you'd certainly expect it to come down somewhere, and that the rock surfaces would be covered. If there was, you wouldn't have any exposed rock surfaces and there are very clean rock surfaces everywhere that we went, and everywhere any of the *Apollo* crews went, at least in the equatorial region. Now, there may be something unknown that sweeps those rock surfaces clean every morning, but I doubt it."

"Everybody is stunned at how little detail we know," says Mendell. "If you're going to actually live and work on the Moon, there are still some really interesting questions that we don't know about." Even simple maps are lacking. Mysterious, uncharted lands still spread across the southern regions. A large wedge of terrain lacks any mineralogical data, and there is not much visual data either. At the south lunar pole, there are still question marks, still mysteries at the most fundamental level. Samples from the Moon all come from fairly close to the equator. Both the Soviet *Lunas* and *Apollo* missions could not carry enough fuel to explore near the poles. When humans return, there will be much new territory to cover. However, before humans venture back to set up a permanent residence on the Moon, some of those questions will need to be answered. That is up to the robots, and they've gotten a good start.

THE ROBOT EXPLORERS

Remote sensing and robotic exploration may well be the key to a successful establishment of a lunar beachhead. Robotic missions to the Moon have plumbed its depths with radar, charted its mineralogy with spectrometers, and mapped its topography with lasers and stereo imaging. NASA's Lunar Reconnaissance Orbiter (LRO) is already in preparation for launch, while India's *Chandrayaan-1* is returning data from lunar orbit right now. Both will shed light on key aspects of the Moon's nature, including the confirmation of the presence of water. The LRO's powerful imaging system will return snapshots of objects the size of a coffee table, making it possible to characterize the Shackleton rim in detail. Mendell comments, "In the very beginning, it was clear that the mapping was very poor, photography was poor, knowledge of the gravity field was sort of okay, but there were three or four basic global parameters that needed to be better defined, and the Lunar Reconnaissance Orbiter was set up to do that."

The suite of six instruments aboard the LRO is designed specifically to pave the way for future human explorers. The camera system, known as the LROC (Lunar Reconnaissance Orbiter Camera), can image objects smaller than a meter and will map potential landing sites. CRaTER (Cosmic Ray Telescope for Effects of Radiation) is designed to analyze radiation levels in the lunar environment, enabling engineers to understand the hazards to long-term human occupation of the Moon. Diviner measures surface and subsurface temperatures, affording critical knowledge for suit and habitat design. The LAMP (Lyman Alpha Mapping Project) uses faint starlight and sky glow to peer into the dark shadowed areas. It will map areas that are permanently shadowed and search for exposed water ice deposits. The LEND (Lunar Exploration Neutron Detector) observes the flow of neutrons from the lunar surface, allowing researchers to tell what components are in the regolith. It will also aid in the search for water, as hydrogen can be a telltale sign of subsurface water. Finally, the Lunar Orbiter Laser Altimeter

Modern Lunar Explorers

In the past two decades, a veritable armada of robot spacecraft have sailed forth to prod and scrutinize the lunar surface from orbit. They include:

Hiten (Japan, launched 1990) The engineering-test spacecraft flew by the Moon ten times before settling into orbit. In addition to micrometeorite detectors, it carried a small satellite, but the tiny craft's transmitter failed before orbit insertion. Hiten was eventually commanded to impact the surface.

Clementine (U. S., 1994) The craft studied lunar mineralogy and mapped altimetry with a laser.

Lunar Prospector (U. S., 1998) spent a year in lunar orbit assembling the most detailed map of components making up the lunar surface. Its neutron spectrometer detected hydrogen leaking from polar craters, implying that deposits of water-ice might be frozen within permanently shadowed craters. At the end of its mission, Prospector was commanded to crash into one of these craters, but no water was detected.

SMART-1 (European, 2003) was the first lunar probe powered by solar electric propulsion (ion-drive). The spacecraft surveyed the lunar surface at high resolution with wider spectrum for eighteen months, focusing on geochemistry and searching for ices at the poles. The craft not only sought out permanently shadowed areas for ice,

but also permanently illuminated highlands for possible future outposts. At the end of its mission, SMART-1 was commanded to impact the South Pole while the astronomical community observed from Earth.

Kaguya (Japan, 2007) This three-ton orbiter carries two microsatellites and 14 instruments to chart the lunar surface and interior. The craft is also transmitting spectacular high definition video of the Moon.

Chang'e 1 is currently returning data from the Moon. The Chinese craft is making 3D images of targeted sites on the Moon, including many of the south polar region. Its instruments will also measure 14 chemical abundances across the lunar surface, and it will chart the depth of lunar regolith, an important data set for future explorers.

China's Chang'e 1 lunar orbiter took this image of the Moon. (photo: NASA/GSFC)

(LOLA) provides precise topographic information for mapping of potential landing sites.

India's *Chandrayaan-1* arrived in lunar orbit on November 8, 2008. It is the Indian Space Research Organization's (ISRO) first spacecraft to leave Earth orbit. *Chandrayaan-1*'s powerful instruments will inventory radioactive isotopes in the lunar surface, helping researches to determine the origin of the Moon.

MORE THAN ORBITERS?

With tightening budgets and a flagging U. S. economy, NASA initially decreed that no more lunar missions beyond the LRO would take place, opting instead for a focus on getting humans there. As Mendell explained, "The robotic program is seen not as a scientific exercise but as a way to gather critical environmental information to enable the engineers to design both the spacecraft and surface systems to go to the Moon. Once you're on the Moon, the presumption is that there will be lots of opportunities to do various kinds of science." In effect, the robotic program was intended to feed the Constellation program. After the LRO, other mission ideas involving rovers or landers were interesting, but NASA managers felt they would not significantly reduce the risk of a human mission. Humans were the priority.

But interest in the Moon has increased in recent years, especially in the arena of robotic spacecraft. European, Asian, and Russian designers have a tremendous amount of capability in robotics, as is being demonstrated by recent lunar spacecraft successes. International partners could supplement and complement the activities that the United States is carrying out, Mendell suggests. "There is an amazing amount of enthusiasm. I really am surprised at the level of the international community's buy-in to the general idea [of a return to the Moon] and the things that are being accomplished."

Paul Spudis, a geologist at the Lunar and Planetary Institute, believes orbiters are not enough. "The upcoming orbital missions should collect a lot of information on the properties and environment of the polar deposits, but we won't really resolve the issues definitively until we go down to the surface and sample and analyze the material in place." In addition to searching for water, robot rovers can test soil strength for landers and habitats, and search out likely candidate sites for solar power and landing areas.

To that end, NASA recently added three small lunar missions. The GRAIL (Gravity Recovery and Interior Laboratory) will orbit twin spacecraft concurrently for several months in 2011, assembling precise maps of the Moon's gravity fields. In effect, GRAIL will do an orbital MRI of the Moon, peering through layers from crust to core.

NASA has also announced the Lunar Atmosphere and Dust Environment Explorer (LADEE), to be launched together with GRAIL. Its hundred-day mission will study the rarefied lunar atmosphere and dust, important arenas for lunar architecture designers. Dust has become the primary concern and driver in lunar designs of future outposts. During *Apollo* surface stays, lunar grit blanketed equipment, coated surfaces inside the LM, and even affected components on space suits such as seals and locking rings. *Apollo 16* Commander John Young experienced the challenge firsthand. He says, "Handling the dust is going to be very difficult. We have to protect our rotating equipment and seals from the dust, or it will stop us."

The LADEE will carry at least two instruments, a spectrometer to study the Moon's thin atmosphere and a dust detector. Researchers hope they will gain insights into suit, habitat, and vehicle designs from the LADEE mission.

Another probe recently added to the cosmic queue is the Lunar Crater Observation and Sensing Satellite, or LaCrOSS. The LaCrOSS will piggyback with the LRO. The mission makes use of the entire upper stage that was used to ferry it and the LRO to the Moon. A small shepherding spacecraft will guide the massive spent booster on a collision course with a permanently shadowed crater floor. As the stage plows into the surface, the shepherd craft will fly through the impact plume, searching for signs of water. The giant stage should provide impressive results. Flight planners expect the impact to send a 1,000 ton plume of material some 70 km into the airless sky. LaCrOSS will sail through the debris, broadcasting real-time images and data about the makeup of the plume. Later, the 700 kg shepherd craft will also be commanded to impact the lunar surface in a site considered a candidate for water ice. Both impacts will be scrutinized by observers on Earth as well as spacecraft in orbit around the Moon.

Over the next decade, planners expect to have enough data in hand to safely establish the first permanent human presence on the Moon, at an outpost that affords humankind real access to a new world. Setting up that historic settlement will require an understanding of the Moon far deeper than we now have. Science is only one of many reasons cited for the need to have a human presence on the Moon, but to many, it tops the list. If applying what science teaches us to everyday life is the ultimate payoff, then science must be the soul of Shackleton outpost. As NASA scientist Wendell Mendell puts it, "It's the Frankenstein problem: you can build the monster, but will it have a soul? Baron von Frankenstein had this tremendous technical achievement where he created life, but there was this little missing bit. It's my job to make sure it has a soul."

THEORETICAL Vs. APPLIED SCIENCE

Science comes in two basic forms: theoretical and applied. It takes theoretical science, the act of doing pure research, to get enough knowledge to apply lessons to everyday life (applied science). At the Moon's first outpost, most early science will be pure research, but historians assert that eventually, this will lead to benefits for all of Earth's peoples.

Wendell Mendell's evolutionary model of planets suggests that the Moon will provide insights into the workings of our own world. Earth and the Moon developed side by side for at least 4 billion years. It will take some time to unravel the skein of scientific yarn, but history shows that the payoff will make the effort worthwhile. Says *Apollo 14*'s Ed Mitchell, "I think of it in the same terms as Antarctica where we spent a goodly portion of the twentieth century just setting up science stations to try to understand the impact of the region on the rest of Earth, weather systems, etc." With their science stations in place, researchers are just beginning to understand the importance of Antarctica to the world's climate, oceans, and atmospheric dynamics, subtleties not imagined a century ago. Mitchell and others wonder what critical insights the Moon will provide. "It's a little different on the Moon, but the idea of a very isolated outpost for exploration and trying to understand the environment is still valid."

If history is any indicator, exploration of space in general, and the Moon in specific, may well play a critical role in humanity's concerns for its own world. A classic example concerns the *Pioneer* Venus mission. *Pioneer* settled into orbit around Venus in 1978. Over the next several years, the spacecraft charted the planet's weather and atmosphere. The craft discovered vast holes in the ozone layer above the Venusian poles. It also discovered the presence of naturally occurring chlorofluorocarbons (CFCs) in the Venusian air. At the same time, many manufacturers in the western world were preparing to debut countless hair sprays, air fresheners, deodorants, and other household products containing CFCs. Carl Sagan and other scientists raised the call that, in light of the ozone state on Venus linked to CFCs, the manufacturers on Earth might want to think twice before pumping the same chemicals into Earth's atmosphere. Thanks to space exploration, an environmental catastrophe may have been avoided. Space historians often cite *Pioneer* as applied science at its finest.

Monitoring Earth's resources from space has become a highlight of the space program, and those studies will undoubtedly continue from the lunar surface. The Moon is better suited than Earth-orbiting satellites for the study of Earth's extended atmosphere and some aspects of the planet's energy fields and particles. With issues of global warming and climate change, space-based Earth studies are taking on more critical roles. "We've had 50 years of NASA," says Goddard Space Flight Center's Jim Garvin, "and if you don't think it has been important, I like to point out that the entire world gave pause when we landed on the Moon, and when we recently landed the rovers on Mars, and when we established an Earth observation system for measuring what is happening to our planet. The causes we don't understand yet, but that's the scientific process. But we're taking the pulse [of Earth], and this is stuff that 30 or 40 years ago would have been considered science fiction."

Pure scientific research often leads to such breakthroughs. But what of the pure science at the Moon? Researchers have first-hand exploration experience and ground samples from only the near-equatorial regions of the Moon. The lunar south pole is a site unlike any *Apollo* landing site, making it a prime scientific goal. Shackleton crater was carved from highland material. The only *Apollo* landing site that is similar is the *Apollo 16* site in the Descartes highlands, but it may differ in ways unknown to science at this time. Remote sensing from *Lunar Orbiter 4 and 5,* along with data from the *Clementine* orbiter, indicate that the southern polar area is similar to the equatorial highlands, composed of gardened and reworked impact breccias (assemblages of angular pieces of rock stuck together by volcanic activity or meteor impacts).

To lunar geologist David Kring, the most important thing about the Shackleton location is its proximity to the South Pole-Aitken impact basin,

Why Go Back? Eight Scientific Reasons

The National Research Council released an Executive Summary outlining the results of a series of meetings among scientists and strategists. The Global Exploration Strategy recommended eight prioritized justifications for a return to the Moon:

1. The bombardment history of the inner Solar System is uniquely revealed on the Moon.

2. The structure and composition of the lunar interior provide fundamental information on the evolution of a differentiated planetary body.

3. Key planetary processes are manifested in the diversity of lunar crustal rocks.

4. The lunar poles are special environments that may bear witness to the volatile flux over the latter part of Solar System history.

5. Lunar volcanism provides a window into the thermal and compositional evolution of the Moon.

6. The Moon is an accessible laboratory for studying the impact processes on a planetary scale.

7. The Moon is a natural laboratory for regolith processes and weathering on anhydrous (lacking water] airless bodies.

8. Processes involved with the atmosphere and dust environment of the Moon are accessible for scientific study while the environment remains in a pristine state.

or SPA. "There are probably a hundred credible sites to visit on the Moon, but South Pole-Aitken is very different from all the sampled sites so far. It's the largest impact basin identified, and Shackleton will have components of the SPA." And if, as Jim Garvin has suggested, the Moon's history is a doorway to understanding Earth, SPA is the place to go.

Kring is not alone in his lust for data from SPA. Dr. Paul Spudis, a geologist at Johns Hopkins University's Applied Physics Laboratory, explains that "its study has the potential to illuminate the earliest parts of Earth-Moon history as well as address fundamental processes of the early Moon." Kring adds, "What happens to the Moon happens to Earth." Hence, insights into the history of the Moon provide insights into the past of our own world.

To gain this much popular momentum for exploration of space beyond Earth, there must be more than perceived scientific and technological value. There is political capital, and it crosses national borders. Even at the height of the Cold War, Soviet and U. S. scientists kept lines of communication open while government leaders gave each other the silent treatment. Soviet scientists invited involvement in such missions as *Phobos 1* and *2,* and the *VEGA* Venus/comet missions. Vachislav Linkin, a scientist at Russia's venerable Institute for Space Research, remembers IKI's connections to the cold-war era western science community fondly. "We felt we could achieve so much together." Today, alliances continue to be forged among nations. The ISS is truly an international endeavor, with Japan's massive KIBO modules, ESA's *Columbus* and *Jules Verne* ATV, and Russia's modules and transport through *Soyuz* and *Progress.* But NASA is hoping to make the lunar outpost more of a partnership than earlier projects. NASA's Wendell Mendell observes, "There's been a lot of conversation with the international community, because on the robotic side the international community has a tremendous amount of capability, as is being demonstrated. They could supplement and complement the things

South Pole Aitken Basin: Visit to a Big Hole

Many researchers would like to see an outpost at the lunar south pole because of its proximity to a huge impact basin called South Pole Aitken (SPA). This gigantic scar may reveal insights into a major issue in modern space science. One of the key theories of planetary development today concerns what scientists call the lunar cataclysm, a cosmic hailstorm of comets and asteroids that ended about 3.9 billion years ago. As planets formed in the early Solar System, they withstood a steady rain of asteroids and comets, the leftovers of Solar System formation. Eventually, planets and moons mopped up most of the cosmic flotsam and jetsam as their gravity pulled in the debris. Craters are the scars left from this violent era, and the largest craters, called impact basins, are the most helpful in determining what went on in those early times. The timeline and details of this Solar System cleanup are not well understood, explains the Lunar and Planetary Institute's David Kring. "In my mind, SPA is the number one site [for future explorers] to visit. The reason is that it's the best place to begin testing a lunar cataclysm hypothesis. We may be able to finally get dates for the end of this period."

Refining those dates is important. Scientists use the biggest scars, the circular impact basins, to estimate these dates. Many of these lava-filled ancient basins, called maria, form the familiar splotches of the "Man in the Moon." The duration of the lunar cataclysm includes all of the basin-forming impacts between (and inclusive of) Mare Nectaris and Mare Orientale. Orientale was the last basin to form. Basins older than Mare Nectaris do exist, but scientists know nothing about their ages. It is usually assumed that they are not part of the cataclysm and formed during an interval that was spread out in time between 4.5 and 4.0 billion years ago (earlier than the most obvious basins we see today).

"When we go back to the Moon," says Kring, "we want to collect samples from those basins to determine their ages. If they all have 3.9- to 4-billion year ages, then they are part of the cataclysm, [which will tell us that it was] 3 times more violent than current estimates." And because our world is so much larger, the rate of impacts on Earth is greater by a factor of at least thirteen.

South Pole-Aitken is the deepest impact basin on the Moon, and undoubtedly holds samples from very ancient lunar history. Its edge is within driving distance from the rim of Shackleton Crater.

Geological map of the massive South Pole-Aitken basin (Clementine geological map courtesy of Paul Spudis, Lunar and Planetary Institute.)

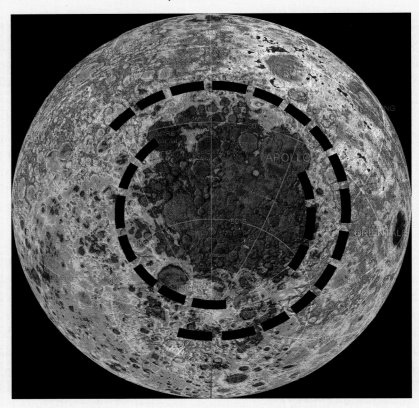

that the United States is doing. Obviously, China has an agenda of its own. India also has an agenda, although they tend to be more cooperative players in the world arena."

NASA's Lunar Systems Project Manager Chris Culbert sees the future of lunar outposts and exploration as having a distinctly international flavor.

Desert Rats

One of the greatest legacies of a lunar program is the development of teams of people who operate on the cutting edge of technology and science. The median age of Project *Apollo* members was 26. Today, a new generation of young engineers and scientists is faced with a set of similar challenges. The Desert Rats program facilitates the interactions of a network of scientists, engineers, academicians, and industry personnel. "It focuses on the lunar architecture and what we're going to need," says Johnson Space Center's Joe Kosmo. "It's been 25 or 30 years, and all the young designers have seen is pre-space activities. Desert Rats is an opportunity for them to see what planetary exploration is really going to mean, in the sense that you're not in a free zero-g environment. You've got gravity to contend with, dust, terrain features. What does a field geologist really

do? What kind of mobility does this person need to do whatever he or she does?" Desert Rats works closely with the U. S. Geological Survey, (or USGS) in Flagstaff, Arizona. It makes use of sites where the astronauts trained for *Apollo*. "It gets young people back into the real world and gets the learning curve going; too much time is spent on the computer and not interacting personally with people, getting into the field, and working things out." The program reflects the strengths of *Apollo*'s culture of self reliance, emphasizing the development of working relationships.

One product of Desert Rats has been the development of a science trailer for use in lunar field work. The trailer carries a rock breaker, microscopic scanner, and fluorescent light for testing samples. Participants transmitted data back to a lab at the USGS as if it was Mission Control on Earth. "Getting dirty!" Kosmo grins. "That's what exploration is going to be all about."

"We need this to be an international endeavor. It may well be that, for example, the Europeans want to push inflatable technology forward, so they're going to have a module that is inflatable, whereas the Japanese want a nice hard shell so they can put this big crane structure on the outside. [NASA's] job is to make sure the architecture accommodates all those different interests."

The first extensive international experience in the human space flight arena came during the joint missions to the Soviet/Russian *Mir* space station.[27] Astronaut Marcia Ivins participated in the program. "There were politics on the *Mir* project, because we were not equal partners. We were a politically mandated guest. We, on both sides, had to suffer the political matchmaking. There were some political things neither side could get around that made the *Mir* flights educational, great experiences, but not equal partnership. So now you move into the Station days, where we are more equal partners across the board. Whatever things were uncomfortable or unsmooth about *Mir* don't exist now. The Russian guys come train with us, and our guys go train over there. The crews are technically multi-lingual, and if your Russian skills or your English skills are not up to par, you don't fly. We were not allowed to do space walks in the *Mir* days, but now the Russians get into U.S. suits, the Americans get into Russian suits. There are still some borders and boundaries that are artifacts of the world's political system, but when you are there with the other two people who are the only ones not on the planet, things are different."

Ivins believes that international involvement as equal partners will be critical to a long-term lunar community. As political relationships heat up between the west and Asia, many see the Moon as common ground for diffusing international tensions.

27. The first joint Soviet/ U. S. mission was carried out in 1975, during the *Apollo-Soyuz* Test Project. Astronauts Tom Stafford, Vance Brand, and Deke Slayton linked up with Alexei Leonov and Valeri Kubasov in the groundbreaking mission.

New Cold Wars, or New Diplomacy?

While building national prestige, space exploration can help us diffuse international crises. During the darkest nights of the Cold War, when political leaders nervously tiptoed through the complexities of international crises, scientists continued an open dialog, and space provided a high ground for combined, peaceful efforts.

Russia has its sights on the Moon again. Their last partially successful planetary probe, *Phobos 2*, reached Mars nearly two decades ago. Now, Moscow's Vernadsky Institute, maker of the successful *Luna* series, plans a return to space with *Luna-Glob* (Russian for "Moon Globe"). This massive probe will continue the search for polar ices and will map the Moon's internal structure, charting the gravity anomalies called mascons.

The Shenzhou VI *lifts off with a crew of Chinese tikonauts. (Photo courtesy of Qin Xian'an, State Satellite Control Center, chief journalist)*

Aside from the two former Cold War superpowers, China is the only spacefaring nation to have orbited humans (Chinese space travelers are called "tikonauts"). Beijing space authorities have played their long-term Moon plans close to the vest. The first decade of the new millennium has been a banner one for the Chinese space program. The orbiting of tikonaut Yang Liwei in 2003 was followed by a flight of two tikonauts just two years later. The Chang'E 1 lunar orbiter has been the only publically announced Chinese lunar mission, but it is billed as the first in a series. The 2,000 kg craft is designed to map minerals and to sense the thickness of the lunar regolith using microwaves. Chinese authorities have referred to an impending lunar rover, and their human space efforts continue to move at a brisk pace. On September 27, 2008, Mission Commander Zhai Zhiqang left the cabin of the *Shenzhou-7* to retrieve an experiment rack from the exterior of the craft. This space walk paves the way for more ambitious missions. *Shenzhou-8* is slated to demonstrate docking, enabling space construction. The Chinese reference to a "small space station" by 2020 has some analysts wondering whether this station will not be circling Earth but rather be perched on the surface of the Moon.

Some analysts have likened the secretive political culture of twenty-first century China to that of the twentieth-century Soviet Union. Several Chinese government officials have attached a great deal of political bravado to their space endeavors, and the Moon is still seen as a high profile political goal. In a decidedly Cold War tone, China's chief scientist of lunar programming, Ouyang Ziyuan said, "Whoever conquers the Moon first will be the first to benefit."[28] The Chinese online news agency China View[29] quoted Ye Peijian—the chief designer of *Chang'e-1*—as saying that China plans to land a probe on the Moon in 2013, perhaps a small rover. But many international policy-makers see the high frontier as a place of opportunity where nations can work together, diffusing political tension as they combine efforts toward common cosmic goals.

In an apparent attempt to avoid another Cold War-style Moon dash,

President George W. Bush couched his return to the Moon speech in inclusive language, calling upon other nations to join in "a journey, not a race." How the Chinese and other world powers will fit into that journey remains to be seen.

Other countries are encouraging international participation as well. India has had a vibrant space program for thirty years, featuring its own launch vehicles. India has designed and lofted its own communications satellites and now manages a fleet of advanced remote sensing satellites to monitor the sub-continent. Its first Moon probe, *Chandrayaan 1,* is a truly international venture.

The European Space Agency's robust space program is now fully enmeshed in human exploration with the flight of their huge *Jules Verne* cargo carrier to the ISS in March 2008. The vehicle is pressurized, and plans are progressing to "man-rate" it.

28. Quoted in the online version of the UK's *Telegraph, Russia Sees Moon Plot in NASA's Plans* by Adrian Blomfield, May 2, 2007.

29. China View, May 18, 2008.

But will politics at home preserve the vision of humans on the Moon and Mars? Popularity of government programs is mercurial, and analysts are asking just how far U. S. taxpayers are willing to go. Is the Moon the only goal? Representative Barney Frank of Massachusetts wanted to find out. His amendment would have barred use of any NASA funds for human missions to Mars. It was soundly defeated. At a Mars Society conference held in Washington D.C. in 2006, NASA's Brian Chase said, "Even though the Frank amendment didn't specifically recommend cutting specific funding, and we aren't spending money directly on a manned mission to Mars at this time, NASA and its allies effectively argued that much of our technology development and lunar planning are precursors to Mars missions." [30]

Going to Mars will take sixty times as long as a trip to the Moon. The expedition will require new technologies only dreamed of today and cost not millions but billions of dollars. If the people of Earth decide that human Mars exploration is a necessary part of their future, be it for survival or pure knowledge, how will it happen?

30. Comments by Brian Chase, NASA's Associate Administrator for Legislative Affairs at the 9th Annual Mars Society convention, Washington D.C., August 3-6, 2006.

Chapter Six

Going to Mars

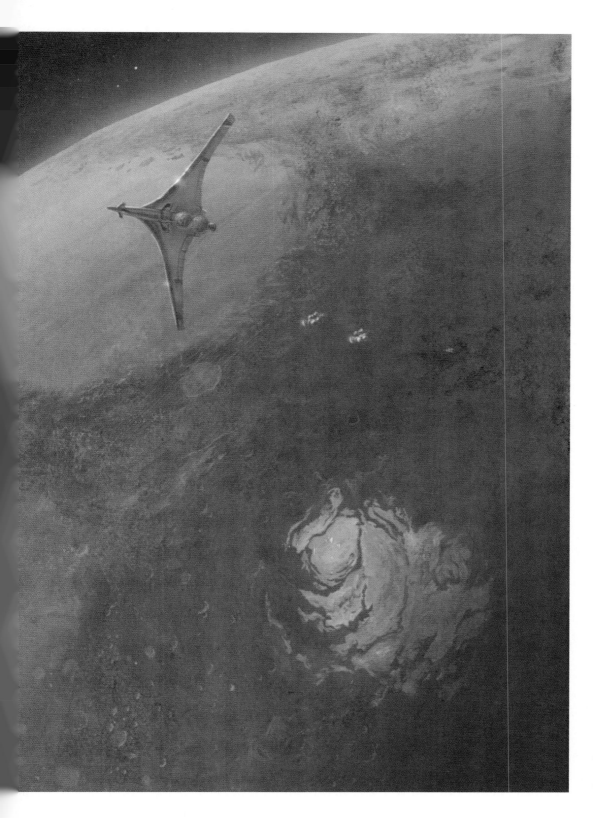

In his 1948 study, Werner Von Braun envisioned several huge winged Mars "boats" taking a crew of 70 to the Red Planet. Below spreads the goal of the first landing boat: the south polar cap. In the distance, we see several of the passenger ships with spherical crew habitats for the long journeys to and from Mars. (Art by author.)

They came seventy strong, in the largest ships ever flown across the interplanetary void. Ten of the 3,700 metric ton vehicles arrived in the vicinity of Mars, dropping into orbit before releasing a massive glider, a "landing boat," to the south polar cap. Skids on the glider secured a smooth landing on the ice, the only sure site for a safe

M. Carroll, *The Seventh Landing*, DOI 10.1007/978-0-387-93881-3_6,
© Springer Science+Business Media, LLC 2009

touchdown. From the pole, expedition members set out in pressurized rovers and supply trailers toward the equator in search of a good landing site. Once an area was chosen, engineers set up camp. They carved a landing strip into the Martian plains, enabling two other landing boats to touch down using wheeled landing gear. While 20 crewmembers remained in orbit, the first fifty "Martians" explored the webbed world. The rest is history.

The armada of ten Mars ships never actually flew. They sailed only in the visions of Werner von Braun, America's preeminent rocket scientist. But the reality will be quite different. Even von Braun, in later years, scaled back his model crew to twelve, realizing that the cost of so much mass in people and equipment was far too dear to cast Marsward. Still, Mars seemed, even then, to be humanity's primary goal for a place to live and work away from Earth. Although the Mars seen through the eyes of modern science is a desolate place, compared to the mysterious red world of von Braun's time, Mars is still the most earthlike place, bar none. It has a day about 37 minutes longer than Earth's, and a similar seasonal cycle, though each season is nearly twice as long (a Martian year lasts about 688 days). Mars is readily accessible in terms of distance and environment. In the past, Mars may have harbored conditions amenable to life, making it a prime target for those searching for insights into life's origins. Most important to human exploration and settlement, Mars has resources for fuel and air, and the most valuable resource, water.

The key to reaching Mars, say aerospace engineers of the twenty-first century, will be infrastructure. Debates rage as how best to lay the path for human landfall on the Red Planet. Should we cache supplies ahead? Make fuel on Mars ahead of human landings? Build cosmic truck-stops along the way? In the days of European expansion across the western United States and Canada, sorties were "flown" by expeditions like those of Lewis & Clarke and John Powell. Once explorers mapped and reconnoitered the frontier, caches of supplies were secured, and trading posts were gradually established. As they were, pioneers built roadways, and transportation progressed from Conestoga wagon to Wells Fargo stagecoach. But it took the roads and the transport to make possible the first true settlements.

In the same manner, the strategy of the next-generation space architects is to forge roads and infrastructure to bridge the chasm between Earth and Mars. Ares V is powerful enough to send large cargoes toward the new world, and skills to live there will be honed on the Moon. But what then? International study groups are springing up across the world to study, in detail, how to establish permanent settlements on the new world using existing hardware or new strategies for "living off the land." But before we settle Mars with shopping malls and indoor parks, it is informative to survey the plans that came early—and often—throughout the space program.

THE FIRST MARS DREAMS

The earliest space-age studies of human Mars exploration may have been done in the Soviet Union. In fact, the original goal for the mighty N-1 booster was not to take humans to the Moon but rather to send a 70 ton unmanned craft to Mars.[31] As early as the late 1950s, Soviet Chief Space Designer Sergei Korolev asked his design team to come up with plans for Mars craft, years before the first cosmonaut achieved orbit. The interplanetary ship, called the TMK, would weigh in at 75 tons and carry a crew of three on a three-year Mars flyby. The craft was to feature 6-meter-diameter living modules that spun for artificial gravity. An instrument module would serve as a storm shelter against radiation.

The U. S. government commissioned its first formal contracts for Mars exploration scenarios in the 1960s, with a few more in the 1970s. The studies were small-scale, backroom affairs carried out by most of the major players in the *Apollo* program: Boeing, Lockheed, Martin/Marietta, North American Rockwell, McDonnell Douglas, and Grumman. In the 1960s, technological limitations made serious study difficult. At about the time of the first Moon landing, then Vice President Spiro Agnew unveiled a NASA proposal to build two titanic nuclear-powered spacecraft. Weighing several hundred tons, the twin behemoths would launch in the winter of 1981, arrive at Mars a year later, and deposit eight astronauts on the surface of Mars for a month. The ships would return to Earth using a high-energy transfer orbit assisted by a close swing-by of Venus. The rusted crew would be home by the summer of 1983. But, as von Braun found in his earlier studies, big ships come

An early Soviet Mars study resulted in this interplanetary Mars craft, the TMK. (Art by author.)

31. This according to Korolev's associate, Vychaslav Filin, published in the Soviet journal *Aviatska y Kosmnavtika*, issue #12, December 1991.

The second Case for Mars resulted in this scenario. A Mars shuttle, descending from a cycling ship, approaches a Mars settlement. Base habitats are buried for protection. Roads connect rows of greenhouses to the inhabited area and landing pads at right. Cooling towers of nuclear power plants rise above a nearby crater rim, above. (Art by the author.)

with big price tags. Additionally, the 1970 social environment suffered from the financial battering of inflationary economies and overseas wars. The plan failed from congressional sticker-shock.

The bad taste left by the studies in the 1970s convinced NASA that Mars was an anathema. Although some plans were quietly considered and discarded, the official line was that Mars was off-limits to human exploration for the foreseeable future. The time had not yet arrived for human Mars missions.

Many in the aerospace industry saw the value of Mars exploration and continued informal work toward the goal of a human presence on the Red Planet. The earliest serious private studies on a large scale were conducted at the University of Colorado in Boulder. There, the first "Case for Mars" conference convened in the summer of 1981. The conference saw such luminaries as aerospace engineer and visionary Robert Zubrin, NASA/Ames planetary scientist Chris Mckay, NASA scientist Carol Stoker, astrobiologist Penny Boston, and many others from industry, government, and the private sector. The energetic group became known as the Mars Underground, partially because of the negative atmosphere evident in official circles concerning human Mars exploration. The Case for Mars II convened in 1984. In attendance was former NASA administrator Thomas Paine, who had formally recommended a Mars plan to the Reagan administration. Three years later, members held the Case for Mars III. Adding to previous Case for Mars work, attendees polished studies, published proceedings, designed concepts such as cycling ships to ferry cargo and crew from Earth to Mars, and fleshed out ways to build Mars settlements.

Three years after the last Case for Mars, astronaut Sally Ride chaired a NASA commission for yet another Mars study. This plan called for two spacecraft, one to carry cargo and one to carry crew. After the cargo ship successfully made it into Martian orbit, the crew would be sent Marsward. After docking with the cargo ship, the astronauts would descend to the surface in a lander, spending two weeks exploring.[32] Flying back to the cargo vessel, the crew would head for home. The proposal came with a massive price tag and was too much for the Reagan administration to swallow.

In 1989, the first President George Bush announced an ambitious, thirty-year program. His Space Exploration Initiative (SEI) included a space station by 1999, a Moon landing a decade later, and humans on Mars by 2019. A NASA study group quickly released the "Ninety Day Report," an analysis of how SEI might be carried out. The resulting scenario called for a 1,000-ton interplanetary craft to be built in orbit at the new space station (which was

---

32. A short stay enables a crew to return to Earth before the planets are out of alignment. Otherwise, the return trip must wait for up to eighteen months.

still ill-defined). Its flight plan was so similar to that of the 1987 study that some complained NASA had simply recycled earlier work. With a price tag of nearly $500 billion, the plan was doomed on Capitol Hill. Congressional reaction was expressed best by a new law that called for an embargo on any funding that might be applied to human Mars initiatives.[33]

Clearly, NASA's large-scale approach to Mars exploration needed to be retooled. That was precisely the goal of Martin Marietta's Robert Zubrin. Zubrin called for a plan that he named Mars Direct. Zubrin's design relied on the automated manufacture of fuel on the Martian surface. A precursor ship would travel, unmanned, to Mars, land, and begin to manufacture fuel. The fuel plant could combine one ton of hydrogen from Earth with elements of carbon and oxygen from the Martian atmosphere to produce 20 tons of propellant for the return trip. The crew would arrive in a ship that carried only enough fuel to get to Mars, as their ticket home would be waiting on the surface. In this creative way, the total mass of a Mars mission dropped from NASA's 600 tons to a mere 88 tons, saving not only weight but money. Instead of earlier plans costing $450 billion, Zubrin's estimates came in below $30 million.

In recent years, NASA has reconsidered Zubrin's Mars Direct, as well as other plans incorporating in-situ fuel manufacturing, creative manufacturing and testing of vehicles, and other cost-cutting strategies. With its incremental approach, NASA managers assert that Constellation incorporates many of the best ideas culled from earlier studies.

NASA is not the only kid on the block. At the Russian aerospace behemoth RKK Energia, studies are under way to build a human Mars mission

Robert Murray's painting of a Mars Direct scenario. (Photo © and courtesy of Mars Society.)

33. Hence, the cancellation of the Transhab project; see Chapter Four.

based on the ISS Zvezda module. The 77-ton craft would carry a crew of up to six to Mars for a cost of $15 billion, depending on international contributions. The proposed mission would last 900 days and calls for solar electric propulsion supported by solar panels that would span seven times the length of a football field.[34]

ALLURE OF THE RED PLANET

The space age stripped Mars of its canals and Barsoomian cities, replacing them with intriguing flood plains, dry lakebeds, towering volcanoes, and precipitous canyons. Rovers continue to send back stunning images of Martian desert scenes that could have been plucked from the pages of an *Arizona Highways* magazine. At some latitudes, water-ice clouds drift over sand dunes and rocky buttes. Mars still stands as the planet most like Earth. In this sense, it offers an appeal that no other world does. But Mars offers something else: a chance to study weather, geology, and climate change in a context outside Earth. Locked within the Martian polar caps is a record of climatic conditions on Mars for perhaps millions of years. This type of data is not available on the dynamic Earth, where rain, wind, and the resurfacing of the planet's crust obliterate most ancient records. Martian climate may reflect changes in Earth's climate, as both planets are subject to changes in fluctuations of the Sun's light and heat, NASA scientist Jim Garvin asserts. "We can understand the evolution of a climate system on another planet so we can better understand our own here on Earth." No other planet has such a record.

In recent years, NASA and ESA have had great success in studying Mars robotically. But robots can only do so much. To underscore this point, NASA/Glenn Mars scientist Geoffrey Landis points to the 1997 Pathfinder lander and rover mission. "When we finished off the mission, someone asked [geologist] Matt Golombeck how long it would have taken a field geologist to carry out all the great science that Pathfinder did in three months. The answer to that question was, 'Probably about an hour.' We've been doing really well with MER[35], but the total distance on each of the rovers averages about 10 km. That's an afternoon's walk for a geologist. The rovers are good, but they're no substitute for a real geologist." Landis likens robotic missions to "sending your friend on vacation and getting them to send picture postcards back. The postcards are great, but it's not the same as being there."

Some of the biggest questions in both science and natural philosophy might be answered by Mars: Are we alone? Is there life anywhere else? If there is life on Mars, did it come from Earth, riding on a meteor blasted from our surface? Or did life on Earth actually *originate* on Mars, making us true Martians? Ames Research Center's Chris McKay says there are two questions that single out Mars as the target for human exploration. "The first question is, did Mars have life in the past, life of a different origin than life on Earth? That's a really neat and interesting scientific and philosophical question. The second question is the question whether Mars could support life in the

---

34. Associated Press, Tuesday January 20, 2004.

35. Mars Exploration Rovers *Opportunity* and *Spirit*.

Alexander Zakharov and the Russian Mars Program

The Russian planetary program was put on hold with the dissolution of the Soviet Union in 1991. But renewed economic strength is reinvigorating Russia's plans for space exploration, and those plans include Mars. Phobos Grunt *(meaning "Phobos Soil") will launch in 2009. The Russian-led international mission is designed to land on Mars's largest moon and return samples to Earth. Alexander Zakharov, chief scientist at the Russian Academy of Science's Space Research Institute, shares his thoughts and plans on upcoming Russian Mars missions.*

The Phobos-Soil mission is very important for Russia. We lost a lot during the last fifteen to twenty years, and this mission has to be a mark of the beginning of reconstruction of space activity for planetary science in Russia. We chose this mission as a very important step for implementation of the next planetary exploration steps, Mars sample return. Phobos-Soil will also study the Martian system, Martian environment, Martian moons, and Mars itself. The Planetary Society's LIFE experiment [which carries microbial life to and from Phobos in a sealed container to test the idea of meteor transport of life from one planet to another] is a very important work. Any bio-experiments are reasons to make anyone nervous. To reduce this nervousness, it is very important to get wide information from well known, famous scientists about goals, science, and experiment details.

I believe it is very important to have international cooperation in human space exploration, not only to reduce expenses for each country or agency but first of all for political reasons. It is difficult to find projects that unite different nations and countries for one goal. Besides, it is essential that a human mission to Mars will be like an Earth delegation. It is an instrument to join different nations for the common goal.

It is important for humans to go to Mars in the future, first of all, due to the human mentality of discoverer and conqueror. For example, the Russians went to discover and then conquer the East—Siberia, and the Europeans to the West—American. Now, the scale of these ambitions is interplanetary, but the nature is the same. The second reason is self-esteem of mankind. We can make this ambitious step. The political aspect is also important. There is competition between nations: Americans, Europeans, Chinese, Russians. I suppose science and rationality in this venture are sometimes at a lower level.

future: Life of its own? Life from Earth? Human life? All of the above? Life on Mars in the past, life on Mars in the future, those are the questions that make Mars an interesting target."

Like McKay, Robert Zubrin sees the reasons for human presence on Mars as going farther and deeper than pure science. "It is a question of fundamental truth. Mars is not just an object of scientific inquiry. It's a world."

THE PROBLEM OF GETTING THERE

Using today's propulsion technology, a trip to Mars, under the best of conditions, takes roughly six months on a fast track. But human missions will be massive, so travel times of up to eight months may be required. Once there, crews will have the option of staying for a few weeks or eighteen months. The reason boils down to planetary alignments. Like Olympic track runners racing in adjacent lanes at different speeds, Earth is constantly overtaking Mars. When a crew arrives at Mars, the return "window," or opportunity to return home, is nearly over. If the crew delays their departure for more than several weeks, they must wait another year and a half for the planets to be in the correct alignment for the journey home. This provides planners with a difficult decision at the start: invest a great deal in resources for a very short mission on the Martian surface, or invest a good deal of time, but run the risks inherent in a long-duration stay. The consensus today is that the first human crew must be outfitted to stay for an eighteen-month expedition. "There's no sense going to Mars for a drive-by" NASA's Jim Garvin maintains. "There's too much to do. It would be the ultimate wasted weekend vacation."

Long travel times pose one of the greatest challenges, and the Center for Space Nuclear Research has a possible solution: nuclear propulsion. Using engines powered by nuclear fission, travel times drop from months to weeks, says Steven Howe, Director of the Center for Space Nuclear Research. The National Research Council formed a committee to evaluate NASA's exploration development program. One aspect of the program is to develop technologies to live on the Moon and travel to Mars. "As part of our briefing, [the Constellation's] Mars architecture team gave us a summary of their results where they looked at fifty different combinations of propulsion and mission architecture. They concluded that the nuclear thermal rocket was the way to go to send humans to Mars. The nuclear thermal rocket has twice the specific impulse[36] of our best shuttle engines." The increased power translates into a Mars craft that weighs half the mass in orbit (nuclear propulsion weighs less). Another use of the extra power is to reduce travel time to Mars. Instead of a nominal 900 day round trip, studies show that a nuclear thermal rocket can make the voyage in 440 days, including a two to three month stay on the surface.

Howe's studies also indicate more efficiency for lunar missions. Using a nuclear-thermal upper stage on the Ares V instead of the conventional one, "we can put 38% more mass on the Moon. If we then estimate what the lunar outpost is going to weigh, something on the order of 250 tons, we save between three and four launches of Ares V. If those are $1.5 billion each, you save $4.5 to $5 billion by just using a nuclear rocket." Current estimates are that developing the stage will take about $3 billion. Howe suggests that while nuclear propulsion is mission enabling for Mars, it is mission enhancing for the Moon as well. "Our argument was that just for the lunar mission, you've paid for the development of this new propulsion technology that now enables you to go to Mars, because you have operational experience by going to the Moon with it, so you can put humans on it now, and you can get much faster missions to the outer planets. It's the technology you need to explore the Solar System."

THE MOON AS TRAINING GROUND

When considering the complexity of going to Mars, EVA manager Glen Lutz comments, "We'll have to get a lot smarter, and we've got missions coming that will get us smarter." Much of the Constellation architecture is based upon the assumption that its technologies, from habitats to electronics to suits and transportation vehicles, will offer direct or indirect lessons for designing such systems for Mars.

NASA Glenn's John Caruso says, "We try to make sure that we're doing [lunar] development that's a leap toward Mars, but there is a significant

---

36. Power of a rocket engine's thrust is measured in units of *specific impulse.*

Mars and Our Future

Along with Apollo 14 *Commander Alan Shepard, Edgar Mitchell took a 3½ mile hike across the plains of Fra Mauro. Mitchell and Shephard spent 9 hours of EVA exploring the highland lunar plateau. Here, he shares a personal view of why Mars is important to humanity's future.*

I think it's vital to put Mars into our future plans. Let me go to a ridiculous extreme: this Solar System isn't going to be around forever. Our Sun is a mainstream star that's about half way through its life-cycle. If our species is to survive, it's going to have to be off this planet. That's a long way in the future, but what it means is that sooner or later we have to become citizens of the universe, and we might as well start working on that right now. We're only now just beginning to get off our planet, much less exploring the rest of the universe or even our own Solar System. Mars is obviously the next best target for that. We're going to have to get off this planet in due course. If we hope to go on to Mars, here is a pretty good place to train and get your techniques down, because it is very certain that we need more training on how to explore and handle this type of hostile environment. What can we learn being on the Moon, practicing there, before you get somewhere several months from home on Mars? It has a lot to offer.

Apollo 14 *LM pilot Edgar Mitchell*

amount of work to just get operational on the Moon. There is a lot that's the same. It's a tough environment, and handling a tough environment on the Moon really does prepare you for handling the environment on Mars."

Constellation's Bret Drake agrees. "One of the biggest challenges for Mars exploration is learning how to live independent of Earth for two and a half years. We've got to have time-on systems. We've got to run things. We've got to understand how they behave over a long, long period of time, longer than the mission duration, so we have a high probability of the system behaving well [during the actual mission]. Mars exploration will be monumental, and it's going to take a lot of work and a lot of data. The way to get that data is to run it and see how it works. The Moon is a perfect place to do that. It stresses the system just right."

NASA's Chris McKay knows first-hand about working in a hostile environment. He has spent the majority of the last two decades of California winters basking in frigid Antarctic summers. McKay believes it is a reasonable strategy to tool up for the hostile lunar environment as a way to safely get to Mars. McKay is now Deputy Program Scientist for Constellation, and his work in the Antarctic wilderness has given him insights into the workings of remote outposts. "If I was building habitats and rovers on Mars, I would be happy to have a team that had done it on the Moon do my design. I wouldn't want to pick up

The Mars Society's Flashline Arctic research station simulates a tuna-can style habitat in the hostile environment of Devon Island in the Canadian Arctic. (Photo by Paul Graham, courtesy of the Mars Society.)

a lunar rover and plop it down on Mars, but using that team to design my rover on Mars would make me feel much more comfortable. It's not just a literal analogy of the hardware; there's the whole human factors and institutional learning that's going to go along with it. Of course the Moon is not Mars, and it's not going to be a carbon copy of what we do, but there's no denying that the experience we gain in building a base and maintaining it on the Moon will be incredibly useful—if not essential—in doing that on Mars."

McKay contends that this is not simply his own opinion. It is the perspective of engineers whose responsibility it is to do advanced design and fabrication of Moon and Mars infrastructure. The engineers in the trenches are the experts, and the opinions of those experts are the ones McKay values most. "It's like when I talk to my mechanic, and he says, 'You need new pads on your brakes.' I have to say, 'Well, he's the mechanic.' I take his word for it, because I believe he understands the workings of the car. So when I went to Johnson Space Center as part of the Lunar Architecture team, it was clear that the systems engineers had come to several conclusions based on their work. One was that 'we think we need to design around the Moon before we design around Mars. It's already hard as hell to design one on the Moon, and it's going to be even harder on Mars.' So I've got to say that these are the people we're relying on to do that, and they're saying to do it on the Moon first. I'm not going to gainsay their opinion. It's well known that scientists love to tell the engineers how to do their work, but at some point you do have to defer to the engineers who are responsible for it. Their opinion, uniformly, is that we need to do it on the Moon first. I think we, the scientists who are interested in Mars, should put a lot of weight on that."

Some Mars-applicable data is accruing at the ISS, says JSC's Brett Drake. "Every time we fly a crew to space station, they're up there 180 days. They go from a gravity environment to zero g and back to a gravity environment 180 days later. That *is* a Mars transfer. It's just like a launch from Earth, transfer to Mars, and landing on Mars. So we're understanding how the human body will perform going to Mars. Also on space station, we're watching how equipment performs. We learn repair techniques. When we get to Mars, we won't have the ability to send replacement parts to fix broken components. We've got to be able to fix things on the fly. These concepts then fold in to when we're on the surface of the Moon." The systems developed for EVA on the Moon eventually apply to the dusty Mars environment. ISS is zero gravity, so it helps engineers plan for the long cruise during Mars transfer. However, Drake says, "you've got to get into a gravity environment, you have to understand how these systems behave in a hostile environment with dust and thermal swings. How systems perform for long periods of time. The thinking is that you can explore and learn and discover a lot about the Moon while you are preparing for Mars."

NASA scientist Jim Garvin adds, "We need the practice to be confident we can go to Mars. The Moon is the obvious place to do that."

The ISS is providing insights and strategies for 6-month human Mars voyages. (Photo by STS 117 shuttle crew courtesy of NASA.)

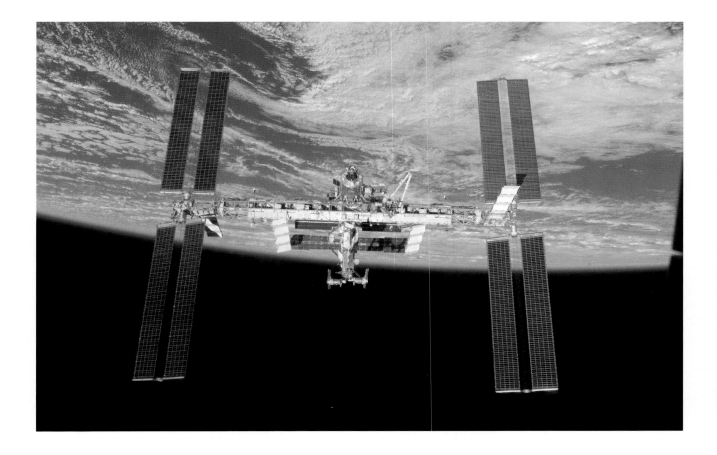

What, specifically, can Mars planners learn from lunar outposts and exploration? Drake has a shopping list, beginning with local taxi service. "Small pressurized rovers: one of the reasons we're going to Mars is to explore and discover. For safety reasons, when we land we'll want to land in safe spots. Unfortunately, those safe spots are generally not the most scientifically interesting. Having a capability to rove long distances routinely is critical. So we'll have small pressurized rovers and we'll run the heck out of them on the Moon, and use those same systems or their cousins on Mars."

Planners also cite the application of robust, continuous power. Power sources on the Moon will play a varied role and will undoubtedly take on many forms. Those systems will inform designers as they plan for Martian exploration and settlement. Having continuous power systems tested on the Moon in its dusty environment will prove some systems and remove others from the list of Mars powerplants.

Communications delays and data forwarding will also teach important lessons. Although the radio signal travel time from Earth to the Moon is just under two seconds, it is up to tens of minutes from Earth to Mars. Delays in communications and handling data in the lunar environment will serve as a less extreme testbed for future Mars mission communications.

Perhaps the greatest lessons to be learned are simply being in a hostile environment for long periods of time. "I'm a big advocate of the need to think in terms of long-term when we go to the Moon and Mars," says Chris McKay. "We should be thinking like we do in Antarctica. The United States has been in Antarctica for fifty years. Just a couple months ago, we did a ribbon-cutting ceremony on a brand new station at the South Pole. The design lifetime of that station is thirty years. We've been there for fifty years and built a station that's designed to last at least another thirty. That's the kind of mentality I would like infused into the Moon program, Moon base, and Mars base. Not that the first mission stays for eighty years, but that we have this view that we're starting this research program that will continue for fifty or a hundred years." NASA has never carried out a program with a lifespan of thirty years. Some analysts feel the Moon offers the perfect proving ground. "If we can't do it on the Moon," McKay observes, "we can't do it on Mars."

Goddard's Jim Garvin agrees that a long-term perspective—with an eye toward Mars as the ultimate goal—is critical in designing lunar systems. "Once we learn to put people on the Moon—whether it's a short period of practicing for deep space access or longer period of sustained permanence—it will prove to us and demonstrate quantitatively what it will take to get to Mars. Much as I want to go to Mars right away, we don't want to make it a joke. It has to be affordable and it has to be sensible."

Garvin also offers a warning: "Remember the number: the amount of mass it takes to send humans to Mars is equivalent to the mass of the completed space station.[37] About 50 to 60 percent of that will be fuel, whether it's

37. Construction on the ISS began in 1998 and is projected to be completed in 2010. The entire structure will require 45 assembly flights.

hydrazine or kerosene or xenon solar electric. The only one that would be smaller would be nuclear, and that's not on the table. These numbers are based on physics, not wishful thinking. We need something to get us ready for a trip like that, and the Moon is the right place to go."

NUTS AND BOLTS OF MARTIAN EXPLORATION

The practical aspects of building a system that enables humans to explore and inhabit Mars are complex. The trip to and from Mars is long and dangerous. Harsh conditions on the Martian surface are ideal contributors to failure of electronics and seals. Constellation Manager Jeff Hanley believes that if Mars plays the pivotal role in informing Constellation designs, the Moon will have a lot to offer. "I think it's *only* critically important to send humans back to the Moon if you intend to go further. So working back from there, how can we inform ourselves along the way? The key to really getting the probability of success—and probability of not killing anybody—sufficiently high is to mature our systems and make them eminently field serviceable. Once you send somebody to Mars, light the rocket and put them on a trans-Mars trajectory, they're gone. There's no changing your mind. There's no turning back. You're gone for a year at least, so the spacecraft must sustain you. We need to get those reliability numbers up. That's an area of technology development we are looking to foster: making systems as robust as possible once you've committed to that long-term outbound trajectory. We're about at the limits of what we can see right now in terms of designing hardy equipment that doesn't break. Now we're shifting our focus to making it eminently field serviceable. Okay, eventually something's going to break. I accept that it's going to break. How do I make standardized components across the system such that I could keep the key systems running to keep me alive?"

Geoffrey Landis feels that efforts can be more focused towards Mars exploration goals. "If I were in charge, the most important thing I would do to get us ready for Mars is to push a little more strongly for manufacturing rocket propellant on the surface (ISRU). It really is the key, not for getting *to* Mars, but for getting *back from* Mars. It's all fine to send robots to Mars and leave them there, but the humans would like to come back, too." Landis feels that propellant manufacturing is the key to enabling humans to travel to and from the Red Planet, and he thinks it needs to be done sooner than later. "What I would like to do is see a sample return mission done with in situ propellant. Just like a human mission, in situ propellant manufacturing should increase the return mass.[38] People are a little bit afraid of it, because it's one new technology to develop for a mission that's pretty hard to do anyway. But if we're serious about going to Mars with people, it has to be done, and given that it has to be done, we should start doing it."

38. This is true because less mass is used for fuel on the return trip. Every ounce of fuel manufactured on the Martian surface equates to an ounce of mass that can be freed up to return samples to Earth. See Robert Zubrin's Mars Direct mission scenario earlier in this chapter.

Living Off the Land

Robert Zubrin is CEO of Pioneer Astronautics. His Mars Direct proposal for human missions to Mars provided a paradigm shift in the aerospace industry, with its use of propellant manufacturing on Mars for the return trip to Earth. Here, Zubrin discusses the use of Martian resources for a permanent human presence there.

When on Mars, do as the Martians will do. When there are people on Mars, will they import their rocket propellant from Earth? No. No sane Martian ever would. The key to the Mars Direct mission is making return propellant on Mars. Of the extraterrestrial destinations available to us, Mars is the most interesting precisely because it has the resources needed to support life and, therefore—potentially—civilization. Why not make use of those resources? The most obvious resource on Mars is the atmosphere. It is 95% CO_2, so there is carbon and oxygen. If you react that with hydrogen, you can produce hydrocarbon fuel plus oxygen. The easiest hydrocarbon fuel to make is methane, and it's a high-performance rocket propellant. In principal, you could get your hydrogen from Mars, too, in the form of water. Even random soil sampled by Viking was 3% water, and *Odyssey* has found regions on Mars where the soil is up to 60%. That's frozen mud—permafrost—in subarctic conditions. But having been to the Arctic, I can tell you that permafrost can be very strong, especially permafrost that's been frozen at an average temperature of −50C. It could be challenging to get the water out. This is why I'm not that thrilled at the concept of getting water out of lunar permafrost at −230C. So what I proposed in Mars Direct was that we simply bring the hydrogen. The hydrogen is only 5% of the total weight of the propellant that results. So the process is this: we land. We run a pump to suck in the air, which is carbon dioxide. We react the carbon dioxide with hydrogen to make methane and water. It's an exothermic reaction, so it releases heat that you can use for various purposes. You then take the water and make hydrogen and oxygen. The hydrogen gets recycled back into the process. The oxygen is a useful product. If you just run this process, you don't get enough oxygen to burn with the methane for fuel, so you split the carbon dioxide into carbon monoxide and oxygen. The carbon monoxide can be discarded. The fuel must be saved cryogenically (refrigerated) so it takes power. [But what you end up with is a methane/oxygen mix for a high performance rocket fuel simply using Mars air and hydrogen that you recycle.] This is all nineteenth century industrial chemistry. For space it has to be made

Experimental hydrogen reduction reactor at Robert Zubrin's engineering company, Pioneer Astronautics. The device brings lunar regolith in via the conveyor at upper right. The funnel to the left guides the regolith into the reactor (brown object at lower center), where hydrogen reacts with lunar iron oxide to produce water and iron.

lightweight, and automated, but there is absolutely nothing new about this chemistry.

Another way one could get oxygen out of CO_2 is with plants. I would not use a greenhouse to make rocket fuel; we need to make oxygen at a much larger rate than that, because for most rockets, oxygen is ¾ to 6/7 the mass of the propellant. But if one was only interested in making oxygen for breathing purposes, breaking down CO_2 through photosynthesis and producing food while making oxygen is a reasonable proposition. It's something that green plants have done for 3.5 billion years.

Assuming that [no hydrogen] leaks, a finite amount of hydrogen can yield an infinite amount of oxygen.

No one expects that; hydrogen can actually leak through solid steel, let alone through joints. But even if you lost one percent of the hydrogen each time, you would still produce eight hundred times the amount of oxygen as the hydrogen you imported.

With the Moon you have a harder situation. There is no water to speak of. There's no air at all. What is there is dirt. What lunar soil consists of, roughly speaking, is something like 10% iron oxides, about 40% silicon dioxide, and around 50% of various oxides including aluminum oxide, magnesium oxide, and calcium oxide. It varies from site to site, of course. The 10% is easy to reduce. The last group is very difficult to reduce. [Using a moderately complex process,] you can get oxygen and carbon monoxide right out of the soil without even bringing along hydrogen. The process is called carbo-thermal reduction. With its different set of resources, the Moon requires much higher temperatures for refinement of this type to work than Mars does. But it's still within the tolerance of many steels. Silicon dioxide cannot be reduced unless you get the temperatures up around 800° C. If it sounds more complicated than hydrogen reduction, it is. But it's doable, and we're working on it. Carbo-thermal reduction also yields silicon. That's the first step in making solar panels.

The in situ process uses heat, and another way of making the refinement of lunar ores more efficient is to recycle that heat. Pioneer is working on a thermal battery to capture some of the heat lost in the reaction, bringing it back to bear on the next cycle of refinement. We're hoping to save 70% of the heat energy.

It is probably the case that the first lunar oxygen-making plants will just use hydrogen. But we will have to eventually move on from there.

DISSENTION IN THE RANKS

Some aerospace analysts and engineers disagree with the current trends at NASA. They argue that the Constellation approach wastes time and money on the Moon—a secondary target—when Mars is the ultimate goal.

Aerospace engineer Robert Zubrin likens the challenge of creating Earth/Mars infrastructure to connecting two posts with rope. "How much rope is needed to connect two posts separated by a distance of 10 meters? It can take any amount, and the rope can be snaked around between the posts. But it can be done with 10 meters if the rope is pulled tight. The issue is whether you want to connect the posts, or whether your goal is to sell rope."

Several alternative scenarios to the Constellation Architecture have been suggested. One approach is to severely cut the number of lunar sortie missions in favor of accelerating human Mars exploration. Another scenario would completely scrap the lunar outpost in favor of sending *Orion/Altair* to asteroids. Supporters argue that asteroid missions would pave the way to early landings on the Martian moons Phobos and Deimos. They propose that flights to asteroids would prove Mars-related technologies more directly than a lunar outpost would. Supporters of the asteroid concept contend that the long-duration flights and tests of heat shields during high-speed Earth return would better simulate conditions during Mars missions. Asteroids are important to study in themselves, proponents say, as they provide windows into Solar System formation and have been the cause of extinctions—and perhaps the bringers of life—to the early Earth.

Private space consultant Doug Stetson, former Solar System program manager at NASA's Jet Propulsion Laboratory, says, "Missions that are interplanetary voyages of six to eight months, to an asteroid, or to the Earth-Sun libration point [a natural parking area in space where gravity from Earth, the Sun, and the Moon keeps an object in one place relative to Earth], may represent significant steps toward an ultimate journey to Mars without all the expense of building a system to live and work on the surface of the Moon. There are good things to do there that have never been done before. I think those things are getting short-changed right now."

Chris McKay shares Stetson's enthusiasm for human asteroid missions. He has been studying such an approach, and how it might fit into the Constellation Architecture, at NASA's Ames Research Center. He feels that an asteroid mission could, in a paradoxical way, speed things up in Mars exploration. McKay believes that what is limiting Mars exploration timelines is support, not some technological breakthrough. "If the public supports an asteroid mission, then that will allow us to develop the capabilities that will be useful in getting us to Mars. It's sort of a Zen problem: the best way of getting to Mars may be by doing other things. I find that in the public there's a lot of interest in missions to potentially hazardous objects. The notion that the world could end gets people's attention. So I think it's an option that should be considered. My view is that Constellation's job is to define options, not make decisions."

The Planetary Society's president, Louis Friedman, is more outspoken on the lunar component of NASA's plans. "If Constellation evolves to a lunar-based goal, it will be a dead end. In the best case, it will be like the space station: We'll get to the Moon and we won't know what to do. In the worst case, it won't even happen, because the American public knows we've been to

Two of the closest views we have of asteroids. Left: 433 Eros, imaged in false color by the Near-Earth Asteroid Rendezvous/Shoemaker craft. From an altitude of 50 km (31 miles), redder hues represent rock and regolith (dirt) that have been altered chemically by exposure to the solar wind and small impacts. Bluer hues represent fresher, less-altered rock and regolith (Photos courtesy of NASA/JPL/ JHUAPL.). Right: The asteroid Itokawa, taken by the Japanese probe Hayabusa, *which landed on the ancient rubble pile in November of 2005 and is attempting to return samples to Earth. (Photos courtesy of ISAS/JAXA.)*

the Moon, and just repeating that over again is not very inspiring. The hope that I have is that Constellation evolves into a sort of international support program for other nations that want to go to the Moon and that we set our sites further. We send astronauts beyond Earth, out to interplanetary space, out to the asteroids, and eventually out to Mars."

Ares launch vehicles have also been brought under scrutiny. An alternate duo of boosters, under study by dissenting engineers at Marshall Space Flight Center, would be easier to build than Ares I and V, they suggest, and would offer more safety and less cost. The alternate boosters are called Jupiter rockets.

In the Constellation scenario, says Wendell Mendell, "there are winners and losers." If the emphasis were to shift away from a lunar outpost and toward a set of asteroid missions, for example, lunar outpost work and Ares V support at Kennedy Space Center would be lost, but increased contracts would be obtained by sites such as Goddard Space Flight Center, the National Oceanic and Atmospheric Administration, and the Jet Propulsion Laboratory. Mendell contends that, "There are people inside NASA who are still waiting for [Constellation] to blow over. There are others who think, wrongly, that their ox has been gored because this new thing appeared on the block, and they think the choices are wrong because the program doesn't do the things they want it to do. They will argue that the things they want to do are fundamentally more important or more interesting or more publicly appealing than the course that has been chosen. What Mike Griffin has done is not the only answer, but it is [well reasoned and] can be executed without breaking the bank."

NASA's Bret Drake sees the bottom line as safety. He believes the safest approach to a multi-year, multi-million-mile Mars journey is to use the Moon as a learning field. "It all comes down to how much risk you are willing to take. Where we are as a culture, failure is not an option. We don't accept failure. If we were willing, as a society, to take more risks and accept the fact that we're going to fail and it's going to happen, the pace might be different, but failure is not an option."

"If you don't ever take risks, it's very hard to do exploration," Geoffrey Landis adds. "As a cultural mindset, that's a problem. We've got to learn to take risks, and when we fail, just understand what's wrong and get back on our feet and keep moving. [NASA is] in a position where Congress and the people who provide funding and direction say, 'Go ahead and do exploration, but whatever you do, don't ever fail.' "

A telling example of current social trends in America is evidenced by the different perceptions of two unrelated flight incidents. The first, the loss of American Airlines flight 587, involved 260 fatalities. The press referred to the incident as a "crash" or "accident." On the other hand, the loss of space shuttle *Columbia*, with 7 fatalities, was called a "disaster." "That's how our society views failure [in the human space program]," says Drake. "We at NASA take crew safety very seriously, and therefore it takes time and a lot of testing in order to make sure that we've got it right."

Robert Zubrin points out that lunar missions are not risk-free. "In an extended series of missions to the Moon, some people will probably be killed. So if your object is saving lives you should, a, skip the Moon, and, b, skip Mars. Furthermore, if your object is saving lives, the money spent on the space program could be much better spent on fire escape inspections, road repairs, child vaccinations, body armor for the troops, and swimming lessons for children. But if your object is to get humans to Mars, then you should send humans to Mars. And if your object is to get humans to Mars while saving as many lives as possible, you should send humans to Mars, skipping the Moon, and using the tens of billions you would have spent on lunar missions on those activities I just mentioned."

Louis Friedman does not believe Constellation should be scrapped in favor of other scenarios. "I don't think we need to lose the progress we've made in Constellation. I do think the basic architecture of getting the Ares V built is a good one. The way to change tracks is to bring in the international partnerships and admit that we don't have the money or rationale to do the lunar base. Delay that, but continue the human space program. You need a set of achievements that go beyond the Moon, and I think that's the way to do it. It would strengthen Constellation as opposed to weakening it."

Historically, pushing into new frontiers has always been a dangerous business. NASA and others who would push those frontiers must continually find a balance between the benefits (discovery, science, human experience, opening frontiers to commercial ventures) versus risk, cost, performance, and schedule.

ENTREPRENEURS AND THE PRIVATE SECTOR

To Constellation project manager Jeff Hanley, the end goal is not only getting people to the Moon and Mars but the elevation of the standard of living across the planet. "If you look at the history of exploration, the resultant improvement in standard of living is tremendous. Look at the trade routes from ancient China throughout Asia and Europe. The first people to blaze those trails were true explorers." In today's world, Hanley believes, the government must fill the role of those early frontier explorers. The *Orion*, which will replace the shuttle and form the backbone of the new Moon initiative, embodies components of both commercial crew transport and cargo delivery to the ISS, Hanley says. "Folks might take a look at that and say, 'What in the world does that have to do with exploration?' To me, it's got everything to do with exploration, because part of the exploration effort must always include—once that trail is blazed—who is to follow." Those who follow the initial lunar and Mars explorers will be the entrepreneurs and pioneers, modern counterparts to those early Chinese traders. "That's part of the whole exploration portfolio. We're trying to create an initial first market, and get players in play."

Communications satellites provide a prime example of this scenario. Initially, these orbiting stations were the product of defense departments. Today, governments across the world buy them from private industry.

Louis Friedman

The president of the Planetary Society talks about motivations behind space exploration and their influence on a human presence on Mars.

Great engineering projects need to be motivated by more than just the desires of those who want to do the building. So many times, the discussion of space ventures is really motivated by people who want to do the work, but they're not the ones who are going to pay for it.

The cathedrals and pyramids were built by people who were thinking about glory and immortality; those were the motivating factors. It wasn't done just for conducting great engineering or architectural projects. The same is true for the fantastic terra cotta warriors of China. Other great engineering projects are motivated by commercial gain, such as the Suez or Panama canals. It's a geopolitical decision to do human missions to Mars. That brings me back to the international cooperation engagement. The space station went nowhere at all for ten years after it was proposed, until it became geopolitically important to engage the Russians in the post-Soviet arms industry to give them something to do. Then, all of a sudden the space station got built. [We need] that same international cooperation type of thinking, to see the importance for the world to get together on a project that engages their highest technologies and their desires for advancing technology and inspiring a new generation. It has to become geopolitically important. That's what it's going to take to get that human Mars commitment. There is a danger, because if we don't do it soon, we'll get more and more satisfied with these robotic missions and maybe people will lose interest in sending people to space. The danger is that there is a trend with being satisfied.

The question of whether humans are hide-bound on Earth or whether they're going to be able to move around the universe is probably going to be determined with Mars. If we can't do it there, it's kind of hopeless. There's nothing better that's in any sense reachable for centuries. And if we *can* do it there, it will

Planetary Society President Louis Friedman

be remarkable because then, within a century of the space age, we can start moving off and living on other planets. So human destiny is really going to be determined by Mars. By the same token, the questions about life originating on other worlds will be, in a large sense, either bolstered by what we find on Mars or not. Mars is unique. It comes down to this: the very fundamental reasons we go into space are going to be determined by Mars. It's not just a candidate place, it's a unique place. It's the only place [where these things] can be determined. Every mission to Mars whets our appetites for doing more.

As CEO of the commercial space 4 Frontiers Corporation, Mark Homnick is one of the entrepreneurs who want to take advantage of those newly blazed trails. "I believe in the government programs laying some of the initial infrastructure for the private commercial space people such as Four Frontiers and the many others, so that we are able to expand. But I think they've already done that over the past decades. For example, back in the sixties [U. S. government space development] was up around 4% of U. S. gross domestic product investment, and that made a tremendous impact on all kinds of ventures. The current investment is down to the $15 to $17 billion level. When we look at the overall space economy, the world GDP for space, it's over $250 billion. So even with the government programs around the world, it still doesn't come to that much. In fact, the commercial space economy has been expanding at over 16% per year over ten years. Commercial space has eclipsed government programs and is developing a life of its own."

Robert Zubrin believes NASA could be more inclusive of private industry within the Constellation Architecture. "I think that the program could, conceivably, be structured so as to have a much more entrepreneurial flavor. For example, if the government commits to a humans-to-Mars program, then they could come along and say, 'We're willing to throw the launch of this open to private competition.' At that point, people like SpaceX (Space Exploration Technologies Corporation) or Lockheed Martin might come along to mobilize and create their own heavy-lift boosters, trans Mars injection stages, a whole bunch of stuff. The thing becomes less contingent on year-to-year funding from Congress. You could even pose it in terms of delivery all the way to Mars. You could say, 'For every 20 tons delivered to the surface of Mars, we'll pay a billion dollars.' That's how we do a lot of things in business."

NASA has sponsored several competitions to encourage development of robotic and human-rated systems. In fact, NASA recently awarded SpaceX Corporation a NASA launch services contract for its Falcon 1 and Falcon 9 launch vehicles. The contract gives NASA the option to order a number of launch vehicles through 2010.

Historically, the entrepreneurial spirit in the space arena has been dampened by cost. 4 Frontiers' Mark Homnick cites the example of mining asteroids, Mars, or the Moon. "From Earth to LEO we would have to get prices down to between $200 and $500 per pound to make it pay. Even with the lowest commercial vehicles, we're an order of magnitude away. However, Yale University has predicted a future lack of copper despite efforts at conservation, starting in about 2030. So, depending on demand, for some rare materials mining on Mars and the NEOs might work commercially." And as space-faring governments or new launch-service providers lay the tracks, Homnick hopes private companies like his will be able to take to the rails. "Think about it: with a Moon program, they're going to put a lot of infrastructure in Earth orbit and on the Moon. That's advantageous to companies like ours."

But Louis Friedman does not think the Moon is sellable to the American paying public in and of itself. "There are two dominant things to keep in mind. The first is the international component. Three Asian countries have lunar programs. Two have announced that they are developing human access to space (one, China, already has). So the twenty-first century perspective is pretty clear. Humans are going back to space beyond Earth. The second thing to consider is the lack of interest on the part of the American public in repeating what's already been done. We must try to do something new and different, namely, the Moon on the way to Mars. We need a new thrust dominated by Mars and by Earth, in terms of monitoring of the environment, developing technology, and so on."

Portrait of a Private Space Corporation

Mark Homnick, CEO of Four Frontiers Corporation, talks about space and commerce.

4 Frontiers was formed on July 20, 2003. Our main focus was the linkage between space technology and space tourism, getting the word out. We felt that was the key element that would overcome this tremendous entry barrier. Generally, getting into space requires a lot of money up front. Technology and other advancements are driving those costs down. Some of the infrastructure that government has put into place is helping, too, but mainly it's the competitive arena that will continue to drive those entry barriers down. We think that's happening all through the commercial space sector. Our technology effort focuses on long-term habitation and material science in space. We believe in people going out into space, going out in a big way, and staying in space permanently.

As for Constellation, any time that multi-billion dollar investments are made in any sector, it certainly has some use. Most of that money is going to what we call "old space," the Boeings and Martin Mariettas and those types of large companies. Very little is going to the commercial startups. For an equal amount of billions invested, in my opinion, we'd get far better results in investing in new commercial space. In general, the competitive sector can do projects on the order of ten times more efficiently than the government. Ultimately, some of it does trickle down, and it keeps the technical folks not only employed but also allows positions for younger folks to shoot for as they're going through their own education.

There is talk about going not only to the Moon but to Mars and the NEOs (Near Earth Objects). Which are most beneficial? I'd have a hard time answering that. They all are beneficial. In an asteroid program, you'd learn a lot about zero-pressure mining. On the other hand, the Martian surface has gravity, atmosphere, and readily available resources. All these things allow what we call MRM (mining, refining, and manufacturing) to be done in a very similar manner to what's being done here on Earth. So the technical [challenges] at Mars are lower than those at the asteroids. The entry, descent, and landing of large payloads is a real challenge, but it's easier to process on the surface than in a vacuum. We get into lively debates as to which of these take precedence, and there are people on both sides of the equation.

From the commercial side, competitive agencies can operate at higher efficiencies, so the private sector could probably do it better and faster once the infrastructure is in place. I would like to see both. When both act in concert, each element does what it's best at. Government is best at putting initial infrastructure in place, especially where it is not yet feasible commercially. Government is a regulating body with rules of the road in place. They operate internationally. They've also done exploration really well. But once there's a path out there, the private sector is good at turning wilderness into home, and all kinds of development springs up. Both are important and have key roles.

Many see the ultimate goal of any Moon outpost as the permanent settlement of Mars. This study, done by 4 Frontiers Corporation, shows a sprawling, nearly self-sufficient Mars settlement. "Tuna-can" habs at center form the beginning of the village, with greenhouses, apartments, and gas factories (behind) added radially from the flanks of a Martian slope in Candor Chasma. Much of the living area would be underground, with light "piped in" from spherical solar shunts along the hillside at right. (© Four Frontiers, art by author)

Space Advocacy Groups:

The Mars Society and the National Space Society

Space advocacy groups argue for various causes in space exploration and settlement. These groups are often involved in education, political lobbying, and support of private-sector space projects. Space advocacy groups such as the Planetary Society, whose founders include astronomer Carl Sagan, or the National Space Society—which boasts 12,000 members worldwide—have successfully saved funding for various space projects threatened by congressional budget cuts. These political lobbies have been carried out on behalf of such missions as the Jupiter *Galileo* probe, the *New Horizons* Pluto mission, and the final Hubble Space Telescope repair mission. Space advocacy groups have also played an important part in the direction taken by NASA's human space programs and ESA's planetary programs. As the private sector takes an increasing role in space development, advocacy groups are gaining in influence and power.

Space advocacy has become important not only in the political arena but in the research field. The Mars Society, founded by Dr. Robert Zubrin, sponsors two Mars habitat simulators, one in the deserts of Utah and one in the Canadian Arctic at Devon Island. These Mars habs were first proposed by NASA/Ames researcher Pascal Lee. Eleven missions have been "flown" at the Flashline facility in the Arctic, some for months at a time in the hostile environment. An impressive 71 mission simulations have run at the Mars Desert Research Station in Utah, outside of Hanksville. Tests have included field science and robotics. A total of 480 people have trained to crew the facilities, including aerospace engineers, scientists, and journalists.

The Mars Society's latest science-related endeavor was its "Mars Project Challenge." The contest guidelines stated that entrants must focus on enabling technologies for a human presence on Mars, and their projects must cost less than one million dollars, including launch costs. The winner, announced at the Society's August 2008 conference, was aerospace engineer Tom Hill's microsatellite TEMPO[3]. The Cubesat-based craft will measure roughly 8 inches long before full deployment. "Cubesat is a 4 by 4 by 4 inch size, and you can triple it," Hill says. "We use one and a half cubes to hold the thruster and electronics, and the other half cube holds the tether." Once in orbit, the tiny satellite will separate into two sections linked by a tether. A simple thruster on one side will start the assembly spinning. Accelerometers will measure the dynamics of the satellite. Integrated accelerometers assure data in all three axes of movement. Hill's invention would rely on amateur high-tech listeners around the globe. "Any satellite member or ham radio operator with a computer hooked up to their receiver tuned to the proper frequency should be able to pick up the carrier signal saying it's our satellite, and then they'd get a stream of data afterwards that we'd ask people to collect from around the world and e-mail to us. We would then know the gravity that was generated at that particular time." The data will be useful in future designs for artificial gravity in long-duration Mars missions.

Geoffrey Landis agrees that Mars is important to future human exploration. "The Moon really shouldn't be the end of exploration; it's really only the beginning. Likewise, even Mars is not the end. Mars is just the next step as we expand out into the Solar System."

THE BIG DECISIONS

NASA is at a crossroads. Destinations in Earth orbit and on the Moon are nearly within reach of private industry. As the agency moves from a near-Earth human presence to space beyond Earth orbit, NASA must redefine its human space program. What directions will NASA ultimately take? Will it boldly incorporate a permanent human presence on the Moon and expeditions to Mars? "The first fifty years of NASA have been transformational," says Jim Garvin. "What are the next fifty going to be? Something? Nothing? Bigger, better? Less? That's all in the hands of the taxpayers and their representatives. We're really at a time of decision. Are we going to enable enhanced future human exploration with robots of this accessible universe or not? Just 40 years ago, for whatever reason, there was a speech that said we are. The Constellation is the first step, as is the Lunar Reconnaissance Orbiter and some of our robotic Mars explorers like Mars Science Laboratory. The question to the public is: Do we want to keep this going, increase it, or flatten it out as a care-taking program?"

The fact is that any U. S. Mars initiative will be done at the level dictated by the American people, Congress, and NASA, who will decide how long people stay, how much practice they need, and what they need to set up and leave behind. As the private sector becomes more involved, space leadership must decide what commercial entrepreneurial involvement is possible, needed, or healthy to incorporate as those colonies on the Moon develop. Garvin suggests several possibilities. "Maybe the Moon becomes something that is a permanent outpost. Maybe it's a sustainable outpost. Maybe it's an occasionally visited outpost or even a commercial outpost. Those transformations are beyond Constellation, but Constellation is a great first step."

The pace of humanity's return to space beyond low Earth orbit will also be dictated by foreign concerns. Will the world's spacefaring nations, in a spirit of cooperation and camaraderie, embrace a world effort to build a permanent presence on the Moon with Mars exploration hot on its heels? What effect will competition have, both diplomatically and commercially?

ONTO MARS?

In 1971, Marvin Gaye recorded the landmark song "Inner City Blues," in which he sang the words, "Rockets. Moon shots. Spend it on the have nots."[39] His song was a call to a society, a message about the prudent use of resources. What should we spend our money on? Many at the time saw the decision simplistically—spend it on the Moon or spend it on the poor. The decision, in reality, is far more nuanced. Will investing in the exploration and settlement of the Moon bring benefits to all people of every social strata and every nation? Can exploration be done in a prudent way so as not to squander precious resources? Will the long-term payoff be worth it? These are the questions that a new spacefaring generation must answer as they face the bold prospect of an international outpost on the Moon.

Says NASA's Glen Lutz, "There's a part of every person that is an explorer. We want to see NASA back in the exploration business. Boldly going where many have been before is not quite as glamorous. But going back to the Moon, we need to get the public to see that it is, in part, a practice run for a mission that's as big an exploration job as any human on the planet has ever done. We need a runway to figure all that out, and that's what we'll get with the lunar experience."

39. *Inner City Blues (Makes Me Wanna Holler),* Words and music by Marvin Gaye and James Nyx. ©1971 (Renewed 1999) JOBETE MU. S.IC CO., INC. All rights reserved. International Copyright Secured. Used by permission.

Constellation's Jeff Hanley sees the implications as wide-ranging. "As a species, as a civilization, I think it is a natural progression for our culture—and I'm not just talking about American culture. It's an imperative of the species to explore, to live on the edge of what is known, and then to reap the benefits. It's all about context to me. The collateral benefits that come out of an exploration effort that were completely unforeseen are striking. There is a many-fold return on exploration, over and over again." Hanley points to the U. S. Exploring Expedition of the 1840s. It was the first large-scale government-sponsored exploration since Lewis and Clarke. The Exploration constituted a several-year mission. "They confirmed the existence of Antarctica; they mapped the South Sea islands so well that those charts were used until the 1950s. They surveyed and mapped the northwest coast of the United States. All the artifacts they brought back at the end of their mission became the Smithsonian Institution. Did they set out to create a world-renowned organization? No. But look at the contributions the Smithsonian has made to world culture."

Others, like Robert Zubrin, assert that Mars is part of a natural progression for a healthy human species. "The Hawaiian islands popped out of the ocean. The birds flew overhead and dropped seeds, and brought life to those places. There is oxygen in the air because life put it there. There is soil on the ground because life put it there. This is what we do. It would be unnatural if humans didn't drop the seeds of life on the islands out there in the cosmos."

The U. S. space agencies involved in project Constellation are stuck between a rocket and hard place. On the one hand, the political machinery—those who hold the purse strings—want NASA to approach a return to the Moon and Mars expeditions in incremental, cost-effective steps over long enough periods, so that the price tag stays below the radar of budget-cutters. Some suggest that the methodical approach reflects a society that does not accept the dangers and risks of space exploration. This approach frustrates the visionary contingent, which would like to see a reasonable but fast return to lunar space and beyond. As Robert Zubrin puts it, "Since when has NASA's job been to be conservative and prudent? NASA was created to storm the heavens in a spirit of challenge!"

Chris McKay counters that, "Bureaucracies are never going to storm the heavens. NASA is a government agency that must operate within regulations. Mike Griffin was once on [the Zubrin side] of the fence, and he used to say things like that, too. When he became head of NASA, reality sank in. Could it be faster? Of course, but I think Mike Griffin has made some good decisions. When Constellation is done, we'll have a transportation system that replaces the shuttle—the Ares I, and the Ares V. The majority of talent and energy at NASA has gone into the Ares vehicles. There are a few working on what we would do on the Moon, and fewer working on what we would do on Mars, and even fewer of us working on asteroids. Those studies are really token. For now, the real effort is building the vehicle, and I think there's some logic there. There's no point in worrying about where we're going to go until we have a vehicle to go with. NASA also realizes that *where* they're going to go is a decision that's going to be made by a future administrator." What the next administration is going to get is the Ares I and V, both in advanced stages of development. Where they go with the new space transports—whether they stick with the nominal plan of going to the Moon, or whether they go to an asteroid or directly to Mars—will not constitute a major change in strategy. The infrastructure will be in place. In short, what Ares and *Orion* will give to future administrations, in a word, is options. "The shuttle is going to retire, and we're going to be flying Ares I and V for the next twenty years," McKay points out. "That decision is behind us. The decision as to what we do with those is clearly ahead of us. It's the right way to do it."

The Constellation program has gotten farther than any other post-*Apollo* Moon program. *Orion*, Ares I and Ares V are enabling technologies. These transportation systems are not chained to any specific destination. Rather, they afford flexible options for exploration of many destinations. Constellation has traction within the industry. It has financial support from both sides of the congressional aisle. It has bipartisan political backing, at least for now. Because of the nature of funding cycles and the momentum of government bureaucracy, Constellation will continue into the foreseeable future. But more importantly, it has put in motion a vast array of international efforts to define and begin to build the strategies and hardware to permanently return to space beyond Earth orbit. When humanity finally does return to the Moon, its arrival may not take the form foreseen by Constellation planners. But after decades of false starts and cancelled programs, the world's spacefaring nations are gaining momentum toward that goal, and Constellation has served as a critical catalyst. Despite varied strategies and directions of study, visionaries the world over are stacking hands in an effort to see humans back in real, exploratory space travel for the first time since 1972. And to many, NASA has regained its vision.

Johnson Space Center's EVA designer Glen Lutz is a hard-working manager. So is Lockheed Martin's Bill Johns, as is Marshall Space Flight Center's Steve Cook. But in their own way, they are visionaries. NASA has many visionaries, people who think outside of the box. So does the European Space Agency, and the Russian Space Agency, the Japanese Aerospace Exploration Agency, and countless others from small businesses to major corporations the world over. These visionaries don't always stand in front of microphones or behind podiums. Some are at work building Styrofoam-and-plywood habitats, knowing that their odd-looking fabrications may one day lead to the first human beachhead on another world. Others spend long hours at computer monitors, formulating the best approach to tool the world's most powerful launch vehicles. Some haunt the back rooms of Johnson Space Center, or Ames, or NASA/Glenn, creating working models of lunar rovers. And astronauts from many countries put their lives on the line to play out the next generation of space exploration aboard vehicles like the shuttle, *Orion,* and *Altair.* To many, it is far more than a job. It is a journey that will continue long after they finish their part. As Lutz put it, "I'll be long retired—hopefully still alive—when we get to Mars, but I want to help them get there with what we're doing now. I'd like to see us capture that exploration piece of people's imagination."

In a very real sense, the prospect of human Mars exploration will play out like the great cathedrals of pre-Renaissance Europe. At a time when civilization consisted of fortresses and low-lying hamlets, massive stone cathedrals rose from the medieval plains of France, England, and Germany. Craftsmen came from hundreds of miles away to contribute their talents as glaziers, carpenters, and stonemasons. Often, construction of these monuments to faith spanned a century or more. Workers knew that their skills would lead to a magnificent structure that only their grandchildren would see. But the work was far more important, far bigger, than any individual. They were building something for the ages, an awe-inspiring edifice to be used and enjoyed for many generations to come. The seventh landing of humans on the Moon—with its promise of worlds beyond—will be, to the twenty-first century, what the cathedrals were to medieval Europe. It is something larger than ourselves, something for the generations to come.

THE FINAL ANALYSIS

A future Mars explorer scales a cliff near the Martian south pole. Many technical, political, and financial cliffs will need to be scaled before she will reach her destination. (Art by author.)

It takes several days, moving at an average of 15,000 miles per hour, for a crew of humans to reach the Moon. It will take six to eight months to make landfall on the rusted sands of Mars, years to voyage to the moons of Jupiter. Like those soaring cathedrals of Europe, such immensities humble us, make us feel small, and put us in our place, while at the same time prodding us to greater things. John F. Kennedy's call to "do these things, not because they are easy, but because they are hard" ripples across five decades to a time when we can, indeed, venture back out into the worlds around us, sculpting a new future for humanity throughout the frontiers of the Moon, Mars, and beyond. We are technologically more capable than we were then. We can explore using less money and doing more things, combining the mighty forces of many nations and creative peoples. We are better at applying lessons learned to our society, our culture, our planet. Twelve Americans trod the dusty plains and mountains of the Moon. They went representing the human species. Perhaps it is time to go back, men and women, Europeans and Asians, Africans and Americans, people of Earth moving out into the immensity of the cosmos. There is inspiration to be had, and humanity needs inspiration. There are concrete benefits to reap, as a world economy grows into a technological society. Perhaps it is time to go back to the Moon, not only for its own sake but also as the hills at the foot of the summit. Beyond our own terrestrial shore, across the ocean of the cosmos, the worlds beckon. Can you hear them?

Afterword

To Boldly Stay

We are on our way back to the Moon, this time to stay. Past experience in space suggests that the hardest part of this new plan may be the "stay." The list of programs related to human exploration that the nations of Earth have thrown away or abandoned is long and sad: the Apollo program, the Saturn V rocket, Skylab, and MIR.

Most recently, for a while it looked like the ISS was going to be abandoned before it was even completed. The notion of staying is not part of our culture of space exploration. Some even argue that NASA should be constantly on the frontier, and thus switching from one destination to another is a feature and not a problem.

I don't agree. The Moon will be our first chance to show we know how to stay in space for the long term.

Not all government exploration programs are doomed to short planning horizons and chronic project shifting. There is one very relevant example of how we—and many other nations—have made long-term scientific exploration a reality. The example is the Antarctic.

The U. S. Antarctic Program has maintained a continuous research program in Antarctica for the past fifty-odd years. The program is operated by a special office within the National Science Foundation—the Office of Polar Programs. Scientists and other federal agencies propose research programs to OPP ranging from astronomy to zoology. There are special programs for teachers, writers, artists, and news reporters. All aspects of the Antarctic Program, both logistics and science, are managed from the same Office at NSF. This is an organizational approach that NASA would do well to emulate with its Moon/Mars program, which is currently fractured into several different directorates.

The commitment to stay in Antarctica is clearly evident in the OPP's planning. The new South Pole station has a design lifetime of 30 years. Fifty years in the past, thirty years in the future; here is a long-term research program we can use as a model for the Moon and Mars.

My ultimate interest is Mars and some big questions about Mars: Was there life on Mars and was it of a different origin from Earth life? Can the future Mars be a place where humans live and work, and can it have a global biosphere? To answer these questions we must establish a long-term research base on Mars and conduct investigations for a generation or two. Once we answer these questions we will then be in a position to decide what we want to do about Mars and life—possibly the focus of work for many generations to come.

It is clear to me that we will not be able to build a long-term research base on Mars if we don't first do it on the Moon. We have engineering, management, and operations lessons to learn. And most of all we have to learn to "stay." We have to learn that the Moon and Mars are not places we visit but places, like Antarctica, where we will stay.

As a start, let's plan for fifty years on the Moon. There will be plenty of scientific research to do. After fifty years in Antarctica there is no shortage of scientific projects proposed each year, and the Moon is at least as interesting as the continent of Antarctica. Indeed, the Moon is an entire world, with its natural complexity waiting to be discovered.

The type of science to be done on the Moon is fundamentally field science. This makes it very different from the International Space Station, which is a constructed laboratory. The difference between field science and laboratory science is key. Laboratory science centers on testing hypotheses, while field science centers on discovering nature. Every year, new things will be discovered on the Moon that will raise new questions and spawn new research.

It is enough that NASA builds a research base on the Moon to support a 50+ year program of field work. NASA does not need to build a gas station, or base its plans on some financial return on investment from mining. Others can play these parts if they prove to be practical—an unlikely prospect, in my view. In Antarctica, the mining of resources has proved impractical and colonization uninteresting. After fifty years there are still only research bases in Antarctica, not mining towns or settlements. However, ecotourism is booming. This could happen on the Moon.

Let's conclude that we don't need an "exit strategy" for the Moon; we need a "permanence strategy." The key to staying on the Moon and also going on to establish a long-term research base on Mars is driving down the operations cost of the Moon base so that it uses no more than about a quarter of NASA's total budget. That is the biggest management and engineering challenge we face as we plan our return to the Moon, this time to stay.

Chris McKay
NASA Ames Research Center
Moffet, California

Chapter Notes

Introduction

xv *By 1965, NASA and private industry Apollo-related personnel had reached a total of 376, 100.*
Arnold S. Levine, *Managing NASA in the Apollo Era* (Washington, DC: NASA SP-4102, 1982), Chapter 4.

Chapter One

4 *…engineers and prescient draftsmen at the British Interplanetary Society drew plans…*
These were first published in the *Journal of the British Interplanetary Society*, January and July, 1939.

6 *From 1959 to 1965, the Soviets and Americans tossed at least 18 robotic space-craft toward the Moon…*
For a fine visual overview, see Don P. Mitchell's website and processed Soviet photographs at www.mentallandscape.com.

11 *Leonov described it as being "like a big loaf of bread…"*
From personal discussions between the author and General Leonov, 1987, Myvatn, Iceland.

11 *In his excellent book,* Two Sides of the Moon, *the hard-working Leonov says…*
This quote is from *Two Sides of the Moon* by David Scott and Alexei Leonov, St. Martin's Press, 2004.

17 Art by Nick Stevens. His website can be found at: http://www.starbase1.co.uk/galleries/index.html

23 *Laboratory technicians would later discover freeze-dried bacteria within the foam insulation of the camera assembly.* From NASA SP-235, *Apollo* 12 Preliminary Science Report.

30 *Analysis indicates that the sample had awaited Jack Schmidt's sample scoop for 109 million years…*
For more on the Tycho age, see Bottke, W. F., Vokrouhlicky, D., and Nesvorny, D. "An asteroid breakup 160 Myr ago as the probable source of the K/T impactor." *Nature* 449 , 48–53 (2007).

Chapter Two

For a nice overview of the Constellation program, see *Project Constellation: Moon, Mars and Beyond* by Tim McElyea (Apogee books, ©2007)

37 *While we might wish that 'off the shelf' EELVs…*
Michael Griffin, remarks at the Space Transportation Association breakfast, January 22, 2008, Washington, D.C. http://www.spaceref.com/news/viewsr.html?pid=26756

42 *At that point, three main parachutes, each 150 feet in diameter, open to carry the booster safely to waiting recovery ships below.* The parachutes are manufactured by Houston's United Space Alliance, which brings decades of shuttle experience to the vehicle.

44 *The most obvious difference is Orion's solar panels, a departure from all previous* U. S. *human-rated spacecraft.* This does not include space stations such as the ISS or *Skylab*. The Soviet and Russian manned programs have used solar panels since the first *Soyuz* in 1967.

Chapter Three

65 *Shackleton's raised rim assures almost uninterrupted contact with Earth, and a near-constant flow of solar energy.* Permanently illuminated lunar highlands were first proposed by the French astronomer Camille Flammarion in 1897. He suggested that some mountains at the lunar poles might be tall enough to qualify as "peaks of eternal light" (*pics de lumière éternelle*).

Chapter Four

83 *I tried [to get NASA to be] more relaxed about walk-back distances.* "Walk-back distance" was the distance NASA estimated astronauts could safely walk back to the LM in case the rover failed. Schmitt and other astronauts wanted their drives to be longer, suggesting that management was too conservative on how far an astronaut could hike in lunar conditions.

87 *Both NASA and Russian experience with long-term space habitation is within the microgravity of Earth orbit.* The only exception is *Apollo*'s LM. *Apollo* astronauts spent less than two weeks' time on the lunar surface, but their lunar "habitat" also served as airlock and spacecraft cabin.

93 *What is it like to hike across a lunar valley, or scale the mountains of the Moon?* From the author's personal interviews with the astronauts.

Chapter Five

101 *In the forty years that spacecraft have been exploring the solar system, researchers have come to understand several common themes that govern the solid worlds.* These worlds include the terrestrial planets (Mercury, Venus, Earth, and Mars), as well as Earth's Moon and the major moons of Jupiter (Io, Europa, Ganymede, and Callisto) and Saturn (Titan). Other solid moons are small enough that they are influenced somewhat differently.

102 *After all, the KT (Cretaceous/Tertiary) impact had an effect on life; it wiped out the dinosaurs.* See Reuters news release *Distant space collision meant doom for dinosaurs* by Will Dunham, Wednesday, September 5, 2007.

115 *The National Research Council released an Executive Summary which outlined the results of a series of meetings among scientists and strategists.* See **The Scientific Context for Exploration of the Moon** by the Committee on the Scientific Context for Exploration of the Moon and Space Studies Board, Division on Engineering and Physical Sciences. ©2007 by the National Academy of Sciences.

119 *India has had a vibrant space program for thirty years…* John Logsdon testimony given before the Senate Committee on Commerce, Science and Transportation; Science, Technology, and Space Hearing: International Space Exploration Program, Tuesday, April 27 2004 - 3:30 PM - SR - 253

Chapter Six

122 *The armada of ten Mars ships…sailed only in the visions of Werner von Braun, America's preeminent rocket scientist.* Von Braun honed many of his later Mars studies at the Marshall Space Flight Center, where he was director and head of the Saturn V program.

123 *…the original goal for the mighty N-1 booster was not to take humans to the Moon but rather to send a 70 ton unmanned craft to Mars…* Filin is quoted in James Harford's excellent book *Korolev* (Wiley & Sons, 1997).

123 *The interplanetary ship, called the TMK…* Published in the corporate history of RKK Energia, 1996. For more on this, see Anatoly Zak's fine website on the history of the Soviet/Russian space program at: http://www.russianspaceweb.com/index.html.

124 *In attendance was former NASA administrator Thomas Paine, who had formally recommended a Mars plan to the Reagan administration.* For a detailed overview of the plan, see *Pioneering the Space Frontier: The Report of the National Commission on Space*, Bantam Books, May 1986.

125 *Instead of earlier plans costing $450 billion, Zubrin's estimates came in below $30 million.* These numbers can be deceptive, however. Zubrin's plan called for two Mars ships, while the SEI included 30 Moon missions and 6 Mars flights.

125 *RKK Energia, studies are under way to build a human Mars mission based on the ISS Zvezda module…*
For more detail, see Energia's website explanation of the mission at http://www.energia.ru/english/energia/mars/concept.html

126 *If there is life on Mars, did it come from Earth, riding on a meteor blasted from our surface?*
This is statistically possible. Scientist James Head estimates that at least one rock from Mars hits Earth every month. (For more, see the Proceedings of the 34th Lunar and Planetary Science Conference, March 17–21, 2003.) Adding to the argument is data first released on August 7, 1996, by a team of NASA scientists implying that the Mars meteorite ALH84001 might contain evidence of ancient life on the Red Planet.

143 *In 1971, Marvin Gaye recorded…*
Inner City Blues (Makes Me Wanna Holler), Words and music by Marvin Gaye and James Nyx. ©1971 (Renewed 1999) JOBETE MU. S.IC CO., INC. All rights reserved. International Copyright Secured. Used by permission.

146 *In a very real sense, the prospect of human Mars exploration will play out like the great cathedrals of pre-Renaissance Europe…*
For an excellent—and historically accurate—description, see Ken Follett's novel *Pillars of the Earth* (NAL Trade, February 2002).

Appendix 1: The Evolving Space Program

Space exploration paints an ever-changing landscape. Launch vehicle decisions shift and morph.

In the middle of 2009, the Obama Administration gave NASA a directive to initiate an independent review of all ongoing U.S. human spaceflight plans and programs, as well as alternatives, which could result in modifications to Constellation launch vehicles. Although former NASA administrator Mike Griffin suggested that the new family of Ares launchers would be less expensive to build than modifying the existing Delta or Atlas boosters, new analyses indicated that the issue is worth revisiting. NASA requested an independent study by the Aerospace Corporation. The goal was to prove that Evolved Expendable Launch Vehicles (EELVs) such as the Atlas V Heavy and Delta IV Heavy were not options. However, the study, released April 2009, showed that both Atlas V and Delta IV have the capacity to carry Orion vehicles to both the International Space Station and lunar orbit.

One objection to the use of EELVs was the existence of "black zones," periods of time during ascent when crews could not safely abort. New studies show that black zones can be avoided by using optimized flight paths.

United Launch Alliance, builders of both the Atlas and Delta launch systems, points out that explosions of liquid propellant used by their launchers are benign compared to solid rocket explosions that would occur early in the flight of Ares 1. Atlas is a proven system, with earlier Atlases launching piloted Mercury capsules. Delta has also accrued an excellent record of safety and reliability. Atlas and Delta expendable launch vehicles have carried nearly 850 combined payloads to space. ULA states that Atlas can be upgraded to human safety levels "with minor modifications." The high-volume flight rate of Atlas for both NASA, the Defense Department, and commercial groups improves mission assurance, ULA says. Delta systems have not been studied in as much detail.

A major drawback with a change from Ares to an EELV concerns NASA's work force. Ares can use facilities and personnel from shuttle-derived programs, creating a smooth transition in both employment and fabrication. But if Constellation changes to an EELV, major job losses and loss of skill sets may be in the offing, NASA sources say. Additionally, four years of Ares development progress would be lost. As Mike Griffin put it in a recent speech, "A fictional space program will always be faster, cheaper, and better than a real one."

Moving from an Ares 1 to an EELV will not solve the ultimate goal of advanced exploration, however. No EELV can be upgraded to carry payloads equivalent to the Altair Lunar Lander, or the equivalent on Mars transfer flights. One NASA source suggests that it might make financial sense to use EELV's for Orion launches and continue development of Ares V for Altair and beyond, spreading the work between programs.

As is true of any long-term, large-budget program, Constellation may have foundational changes on the long road to worlds beyond.

The following paintings, all by the author, show a little of how ideas of exploration have changed over the years.

Moon Base: Early concept for a solar power tower at the lunar south pole. The height of the solar plant would keep it in sunlight longer than surface-mounted panels. Note the ice in the foreground.

Outpost: A 1990 vision of a lunar outpost on the Moon's limb in the crater Grimalde.

Near-Earth asteroids are important targets for Constellation-enabled crews.

Constellation technology will ultimately enable humans to explore Mars. This crew is visiting the historic site of the second Mars Viking lander. (Shown courtesy of Four Frontiers)

One pre-Constellation concept for Mars travel involved cycling ships such as this one, presented at the Case for Mars conferences. (Based on sketches by Carter Emmart)

Once a settlement is established, Mars exploration will take on many forms. Here, a Mars blimp observes an incoming spacecraft. (Shown courtesy of Four Frontiers)

Appendix 2: Moon Missions

Spacecraft	Goals/Results/Achievements
Pioneer 0 (US)	Lunar orbit attempt; launch failure 8/17/1958
Luna (USSR)	Lunar impact attempt; launch failure 9/23/1958
Pioneer 1 (US)	Orbit attempt; launch failure 10/11/1958
Luna (USSR)	Lunar impact attempt; launch failure 10/12/1958
Pioneer 2 (US)	Lunar orbit attempt; launch failure 11/8/1958
Luna (USSR)	Lunar impact attempt; launch failure 2/4/1958
Pioneer 3 (US)	Lunar flyby attempt; launch failure 12/6/1958
Luna 1 (USSR)	Lunar impact attempt. Launched 1/2/1959. Flyby at 5,000 km range
Pioneer 4 (US)	Launched 3/3/1959. Lunar flyby at 60,500 km; radiation data returned
Luna (USSR)	Lunar impact attempt; launch failure 7/18/1959
Luna 2 (USSR)	Launched 9/14/1959. First lunar landfall; confirmed existence of solar wind
Pioneer Orbiter (US)	Exploded on the launch pad during test 9/24/1958
Luna 3 (USSR)	Lunar flyby 10/06/1959; first images of the Moon's far side
Pioneer Orbiter (US)	Launched 11/26/1959. Shroud failure necessitated booster destruction
Luna (USSR)	Lunar flyby attempt; launch failure 4/15/1960
Luna (USSR)	Lunar flyby attempt; launch failure 4/16/1960
Pioneer Orbiter (US)	Lunar orbit attempt; launch failure 09/25/1960
Pioneer Orbiter (US)	Lunar orbit attempt; launch failure 12/15/1960
Ranger 1 (US)	Launched 8/23/1961. Engineering flight test, partially successful
Ranger 2 (US)	Launched 11/18/1961. Engineering flight test, partially successful
Ranger 3 (US)	Launched 1/26/1962. Hard landing attempt; missed Moon by 36,808 km
Ranger 4 (US)	Launched 4/23/62. Hard landing attempt; missed target, impact on far side
Ranger 5 (US)	Launched 10/18/1962. Hard landing attempt; power failure

Sputnik 25 (USSR)	Soft-landing attempt; launch failure 1/4/1963
Luna (USSR)	Soft-landing attempt; launch failure 2/3/1963
Luna 4 (USSR)	Launched 4/2/1963. Soft landing attempt; missed Moon by 8,500 km
Ranger 6 (US)	Launched 1/30/1964. Attempted close-ups before impact, no images returned
Luna (USSR)	Soft-landing attempt; launch failure 3/21/1964
Luna (USSR)	Soft-landing attempt; launch failure 04/20/1964
Zond (USSR)	Probable lunar test of Mars probe; launch failure 6/4/1964
Ranger 7 (US)	Lunar imaging 07/31/1964. Spacecraft returned 4,308 images before planned impact
Ranger 8 (US)	Lunar imaging 02/17/65. Spacecraft returned 7,137 images before planned impact
Cosmos 60 (USSR)	Soft-landing attempt; launch failure 3/12/1965
Ranger 9 (US)	Lunar imaging 03/21/1965; spacecraft returned 5,814 images before planned impact
Luna (USSR)	Soft-landing attempt; launch failure 4/10/1965
Luna 5 USSR)	Launched 5/9/1965. Soft-landing attempt; crashed in the Sea of Clouds. (According to Soviet authorities at the time, the impact occurred at 31º S, 8º W. However, further analysis suggests the crash site is at 8º N, 23º W.)
Luna 6 (USSR)	Launched 6/8/1965. Soft-landing attempt; missed Moon by 160,000 km
Zond 3 (USSR)	Lunar flyby 07/20/1965. Returned images of far side, flew test Mars trajectory
Luna 7 (USSR)	Launched 10/04/1965. Soft-landing attempt. Crashed west of the crater Kepler
Luna 8 (USSR)	Launched 12/03/1965. Soft-landing attempt. Punctured landing air bag, which caused crash in Sea of Storms
Luna 9 (USSR)	First successful lunar landing (1/31/1966); returned panoramas and radiation data for four days
Cosmos 111 (USSR)	Lunar orbit attempt; launch failure 3/1/1966
Luna 10 (USSR)	First successful lunar orbit (4/4/1966); data returned for 56 days
Luna (USSR)	Lunar orbit attempt; launch failure 4/30/1966
Surveyor 1 (US)	First powered landing (6/2/1966). Returned 11,237 images over six weeks
Explorer 33 (US)	Launched 7/1/1966. Lunar orbit attempt; stranded in Earth orbit, but returned radiation data
Lunar Orbiter 1 (US)	Lunar orbit 8/14/1966. Returned 229 images taken over eleven days
Luna 11 (USSR)	Lunar orbit 8/27/1966. Lunar gravity and radiation data returned (imaging failed)

Surveyor 2 (US)	Launched 9/20/1966. Soft landing attempt; crashed southeast of crater Copernicus
Luna 12 (USSR)	Lunar orbit 10/25/1966. Orbital imaging during 602 orbits
Lunar Orbiter 2 (US)	Lunar orbit 11/09/1966. Orbital imaging, 817 photos
Luna 13 (USSR)	Soft landing (12/24/1966); returned images plus radiation and soil data
Lunar Orbiter 3 (US)	Lunar orbit 2/8/1967. Orbital imaging, 626 photos
Cosmos 146 (USSR)	Zond craft, launch failure 3/10/1967
Cosmos 154 (USSR)	Zond craft, launch failure 4/8/1967
Surveyor 3 (US)	Landed 4/20/1967. Returned 6,315 images and used soil scoop; visited by Apollo 12
Lunar Orbiter 4 (US)	Lunar orbit 5/7/1967. Returned 546 images
Surveyor 4 (US)	Launched 7/14/1967; crash landing
Explorer 35 (US)	Lunar orbit 7/22/1967. Studied lunar fields and particles
Lunar Orbiter 5 (US)	Lunar orbit 8/1/1967. Returned 844 images
Surveyor 5 (US)	Landed 9/11/1967. Returned over 18,000 images and analyzed soil chemistry
Zond (USSR)	Zond craft, launch failure 9/28/1967
Surveyor 6	Landed 11/10/1967. Returned images and analyzed soil chemistry
Zond (USSR)	Zond craft, launch failure 11/22/1967
Surveyor 7	Landed 1/10/1968. Returned images from the rim of Tycho crater
Luna (USSR)	Lunar orbit attempt; launch failure 2/7/1968
Zond 4 (USSR)	Launched 3/2/1968. Engineering test resulting in solar orbit
Luna 14 (USSR)	Lunar orbit 4/10/1968. Orbital photography
Zond (USSR)	Zond craft, launch failure 2/7/1968
Zond 5 (USSR)	Launched 9/22/1968. Circumlunar flight, Earth return September 21 with living payload of turtles, worms, plants, and bacteria
Zond 6 (USSR)	Launched 11/10/1968. Circumlunar flyby and return to Soviet Union
Zond (USSR)	Zond craft, launch failure 1/20/1969
Luna (USSR)	Lunar rover attempt; launch failure 2/19/1969
Zond L1S1 (USSR)	N-1/Zond test launch attempt; launch failure 2/21/1969
Apollo 11 (US)	First human lunar landing (7/20/1969) at Sea of Tranquility
Luna 15 (USSR)	Launched 7/13/1969. Entered orbit July 17. Sample return attempt; crash landing
Zond 7 (USSR)	Launched 8/7/1969. Circumlunar flyby and return to Soviet Union
Cosmos 300 (USSR)	Rover or sample return attempt; launch failure 9/23/1969
Cosmos 305 (USSR)	Rover or sample return attempt; launch failure 10/22/1969
Apollo 12 (US)	Landing 11/19/1969. Second human lunar landing at Surveyor 3 site
Luna (USSR)	Sample return attempt; launch failure 2/6/1970

Luna (USSR)	Sample return attempt; launch failure 2/19/1970
Apollo 13 (US)	Launched 4/11/1970. Near-fatal in-flight explosion; safe return April 17
Luna 16 (USSR)	Launched 9/12/1970. First successful unmanned lunar sample return mission
Zond 8 (USSR)	Launched 10/20/1970. Circumlunar flyby and return to Indian Ocean
Luna 17 (USSR)	Launched 11/10/1970. First robotic lunar rover, traveled 10.5 km
Apollo 14 (US)	Launched 2/5/1971. Third human lunar expedition (to Fra Mauro)
Apollo 15 (US)	Launched 7/30/1971. Fourth human lunar expedition to Hadley Rille/ Apennine Mountains
Luna 18 (USSR)	Launched 9/2/1971. Sample return attempt; crash landing
Luna 19 (USSR)	Lunar orbit 10/2/1971. In addition to 4,000 orbits, probe's solar wind studies were coordinated with Mars 2 and 3 and Veneras 7 and 8
Luna 20 (USSR)	Landed 2/17/1972. Sample return from lunar highlands
Apollo 16 (US)	Launched 4/21/1972. Sixth human Moon expedition, studying the southern highland region of Descartes
Soyuz L-3 (USSR)	N-1 test with lunar orbiter; launch failure 11/23/1972
Apollo 17 (US)	Launched 12/11/1972. Last human expedition to the Moon, carrying the first scientist, Harrison Schmitt
Luna 21 (USSR)	Landed 1/15/1973. Second Lunokhod rover
Explorer 49 (US)	Launched 6/10/1973. Radio astronomy from lunar orbit
Luna 22 (USSR)	Launched 5/29/1974. Lunar orbiter with extensive orbital changes
Luna 23 (USSR)	Landed 11/6/1974. Damaged during landing, unable to return deep core samples from Sea of Crises
Luna (USSR)	Sample return attempt; launch failure 10/16/1975
Luna 24 (USSR)	Landed 8/18/1976. Deep core sample return from Sea of Crises
Hiten (Japan)	Launched 1/24/1990. Carried out ten lunar flybys, went into orbit 10/2/1991 before impacting; subsatellite Hagoromo deployed to lunar orbit, transmitter failed
Clementine (US)	Lunar orbit 2/21/1994. Lunar surface mapper; fuel loss resulted in canceled asteroid encounter
Lunar Prospector (US)	Lunar orbit 1/11/1998. Mineral mapping; discovered possible evidence of water ice at lunar poles
SMART-1 (ESA)	Lunar orbit 11/15/2004. Europe's first Moon mission, using solar electric propulsion
Kaguya (JAPAN)	Lunar orbit 10/3/2007. HDTV imaging of Moon; studying lunar origins and evolution
Chang'e 1 (China)	Lunar orbit 11/5/2007. 3D imaging, elemental mapping, and studies of soil depth
Chandrayaan 1 (India)	Lunar orbit 11/8/2008. Detailed mapping in visible, infrared, and X-ray

Appendix 3: Mars and Asteroid/Comet Explorers

The ultimate goal of the Constellation Program is to provide an infrastructure that will enable humans to travel beyond Earth. Future exploration targets may include Mars, asteroids, and comets. Below is a listing of spacecraft designed to study these sites.

Mars1960A (USSR)	Possible Mars flyby attempt; launch failure 10/10/1960
Mars1960B (USSR)	Mars flyby attempt; launch failure 10/14/1960
Sputnik 22 (USSR)	Mars flyby attempt; launch failure 10/24/1962
Mars 1 (USSR)	Launched 11/1/1962. Mars flyby attempt; interplanetary studies carried out; contact lost 3/21/1963
Sputnik 24 (USSR)	Launched 11/4/1962. Attempted Mars landing; upper stage failure; reentry 1/19/1963
Mariner 3 (US)	Launched 11/5/1964. Attempted Mars flyby; shroud failure
Mariner 4 (US)	Launched 11/28/1964. First successful Mars flyby July 15 and 16, 1965; 22 photos and other data received
Zond 2 (USSR)	Launched 11/30/1964. Flyby/landing attempt; contact lost May 1965
Zond 3 (USSR)	Lunar flyby 7/20/1965. Returned images of Moon's far side, flew test Mars trajectory
Mariner 6 (US)	Launched 2/26/1969. Second successful Mars flyby July 31; 75 images and other data returned
Mariner 7 (US)	Launched 3/27/1969. Third Mars flyby; 126 images of southern hemisphere
Mars1969A (USSR)	Flyby attempt; launch failure 3/27/1969
Mars1969B (USSR)	Flyby attempt; launch failure 4/2/1969
Mariner 8 (US)	Flyby attempt; launch failure 5/8/1971
Cosmos 419 (USSR)	Launched 5/10/1971. Attempted orbiter/lander; upper stage failure
Mars 2 (USSR)	Launched 5/19/1971. Successful Mars orbiter (achieved November 27); lander failed

Mars 3 (USSR)	Launched 5/28/1971. Successful orbiter (December 2); lander touched down 12/2/1971 but ceased transmission after just 14 seconds. A tethered rover was also aboard.
Mariner 9 (US)	Launched 5/30/1971. First orbiter of another planet: Mars orbit achieved November 14; global mapping continued until 10/27/1972
Mars 4 (USSR)	Launched 7/21/1973. Attempted orbit; retros failed. Flyby on 2/10/1974 returned some data
Mars 5 (USSR)	Launched 7/25/1973. Achieved orbit 2/12/1974; transmitter failure after 22 orbits
Mars 6 (USSR)	Launched 8/5/1973. Lander attempt; contact lost at landing; first in situ measurements of the Martian atmosphere
Mars 7 (USSR)	Launched 8/9/1973. Attempted lander; premature separation cause craft to miss planet; flyby on 3/9/1974
Viking 1 (US)	Launched 8/20/1975. Orbiter global mapping/first successful Mars landing 7/20/1976; operated on surface for over 5 years
Viking 2 (US)	Launched 9/9/1975. Orbital global mapping/successful Mars landing 9/3/1976; surface operations for $3\frac{1}{2}$ years
Phobos 1 (USSR)	Launched 7/7/1988. Attempted orbiter and two Phobos landings; contact lost en route
Phobos 2 (USSR)	Launched 7/12/1988. Mars orbiter; attempted Phobos landings; orbiter failed just prior to final encounter with Phobos before landers could be deployed
Mars Observer (US)	Launched 9/25/1992. Mars orbit attempt; contact lost prior to orbit insertion, probably due to a ruptured fuel line
Mars Global Surveyor (US)	Launched 11/7/1996 Mars orbiter; first use of aerobraking for planetary orbit; lander relay
Mars 96 (USSR)	Launched 11/16/1996. Attempted orbiter and multiple landers; launch failure
Mars Pathfinder (US)	Launched 12/4/1996. Air-bag equipped lander and rover; mission lasted three months
Nozomi (Japan)	Launched 7/3/1998. Orbit attempt, abandoned due to fuel loss; Mars flyby 12/14/2003
Deep Space 1 (US)	Launched 10/24/1998. Encountered asteroid Braille and comet Borelly; solar propulsion
Mars Climate Orbiter (US)	Launched 12/11/1998. Orbit attempt; spacecraft burned up in Martian atmosphere due to a navigational error
Mars Polar Lander/Deep	Launched 1/3/1999. Attempted lander and penetrators; crash probably due to Space 2 (US) premature engine cutoff from software problem
Stardust (US)	Launched 2/7/1999. Encountered Comet Wild2 and returned samples to Earth; en route to Comet Temple 1
NEAR (US)	Launched 12/20/1999. Orbited asteroid Eros; landed on 2/12/2001

Mars Odyssey (US)	Launched 3/7/2001. Orbiter reached Mars 10/23/2001; lander relay
CONTOUR (US)	Launched 7/3/2002. Attempt at dual comet flyby; apparent explosion of upper stage
Mars Express (ESA)	Launched 6/2/2003. Orbiter and attempted landing Beagle 2; orbit achieved 12/25; international lander relay; communications with Beagle lander lost before landing
Spirit (MER-A) (US)	Launched 6/10/2003. Mars rover, airbag landing; has explored over 7.7 km of Gusev Crater/Columbia Hills since landing 1/3/2004
Opportunity (MER-B) (US)	Launched 7/7/2003. Mars rover, airbag landing; has explored over 15.7 km of the Meridiani plains since landing 1/25/04
Deep Impact	Launched 1/12/2005. Flyby of Comet Temple 1 7/4/2005; observed spectra from impactor
Mars Reconnaisance	Launched 8/10/2005. Mars orbit 3/10/2006; high resolution imagery for future landing orbiter (US) sites; search for evidence of present and past water
Hayabusa (Japan)	Landed 11/26/2005 on surface of asteroid Itokawa. (Secondary lander MINERVA failed.); Currently en route to Earth with samples; ETA June of 2010
Phoenix	Launched 8/4/07. Landed near north pole; mission from 5/25/2007-11/2/2007
Dawn (US)	Launched 9/27/2007. En route to orbit asteroids Vesta (from August 2011 to May 2012) and Ceres (in February 2015)
Rosetta (ESA)	Launched 3/2/2004. Encountered asteroid Steins en route to land on Comet Churyumov-Gerasimenko in 2014

Index

A

Accretion disk, 99
ACTS lunar lander, 58
Airbags, 3, 7, 8, 44, 52, 53
Aldrin, Edwin "Buzz", 22, 23
Altair Moon Lander, 44, 54
    Cargo Variant, 57–59
    Outpost Variant, 56–59
    Sortie Variant, 56, 63
Ambrose, Rob, 82
Ames Research Center, 126, 136
Amundsen, Roald, 67
Anders, William, 19
Anorthosite, 29
Antarctica, 64, 88, 89, 96, 113, 132, 144
Apollo, 9–15, 18, 20, 22, 26, 28–31, 34, 36, 40, 41, 42, 44, 46, 48, 50,
    52, 55, 56, 59, 60, 68, 69, 70, 73, 76, 83, 91, 94, 96, 104, 105,
    107, 109, 110, 114, 117, 123, 145
    Apollo 1 fire, 13
    Apollo 7, 18
    Apollo 8, 18, 19, 20, 103
    Apollo 9, 19, 20
    Apollo 10, 20
    Apollo 11, 21, 22, 23, 24, 27, 29, 31, 60, 64
    Apollo 12, 22, 23, 24, 29, 31, 64, 80, 93, 150
    Apollo 13, 24, 25, 26, 27, 28, 83
    Apollo 14, 28, 31, 59, 64, 74, 93, 108, 113, 129
    Apollo 15, 28, 29, 31, 64
    Apollo 16, 29, 31, 36, 56, 64, 72, 93, 112, 114
    Apollo 17, 29, 30, 31, 35, 49, 64, 83, 93, 96, 104, 106
    Command module, 10, 14, 20, 26, 45, 49
    Service module, 10, 18, 20, 25, 26, 27, 44, 54
Apollo Soyuz Test Project, 73, 117
Ares I, 37–43, 54, 137, 144, 145
Ares V, 33, 38, 39, 42, 44, 60, 85, 122, 128, 137,
    138, 144, 145
Ares reusability, 42
Ariane V, 55
Armstrong, Neil, 22, 23, 24
Asteroids, 30, 36, 100, 101, 102, 105, 116, 135, 136,
    137, 140, 141, 144, 145
ATHLETE vehicle, 63, 79, 82, 83, 84, 89, 95
ATK, 41, 42
Atlas V launch vehicle, 37, 38

B

Baikonur, 17, 18, 21, 27, 28, 55
Bean, Alan, 22, 23, 24, 25, 80, 93
Boeing, 38, 42, 60, 86, 123, 141
Borman, Frank, 18, 19
Bosch, Michael, 55
Boston, Penny, 124
Bussey, Ben, 64

C

Caruso, John, 66, 68, 84, 128
Case for Mars, 124
Cataclysm, 102, 105, 116
Cernan, Gene, 20, 30, 35
Chaffee, Roger, 12
Champollion, Francois, 99
Chandrayaan-1India, 110, 111
Chang'e 1China, 111, 118
Clementine lunar orbiter, 65
Cold War, 4, 5, 6, 115, 118
Collins, Michael, 22
Collisions, 30, 66, 101, 102, 113
Columbus Space Station Module, 58
Comets, 36, 66, 100, 101, 104, 115, 116
Communications, 8, 10, 12, 19, 46–47, 48, 57, 58, 61, 63, 83, 92, 94,
    96, 132, 138–139
Conrad, Pete, 22, 23, 24
Constellation, 4, 9, 15, 19, 27, 34–35, 40, 41, 42, 43, 54, 55, 61, 64,
    74, 75, 76, 84, 102, 112, 125, 129–130, 133, 137, 138, 141,
    143, 144–145
Constellation Architecture, 4, 6, 10, 55, 94, 128, 135, 136, 140
Construction techniques, lunar, 96
Cook, Steve, 37, 38, 39, 42, 65, 145
Cosmic rays, 91, 110
Crew Exploration Vehicle (CEV), 4, 40, 42, 43, 48, 50, 52, 53, 54, 58
Culbert, Chris, 61, 68, 82, 89, 90, 92, 94, 98, 116
Cunningham, Walt, 18

D

Delay, Tom, 8, 12, 29, 35, 54, 55, 127, 132, 138
Delta IV launch vehicle, 37, 38, 39, 42
Desert Rats, 117

Devon Island (Canadian arctic), 130, 142
DEXTRE, 80
Diamandis, Peter, 59
Drake, Brett, 61, 129, 131, 132, 137
Duke, Charles, 36, 48, 56, 93

E
Earth Departure Stage (EDS), 37, 38, 44
Earth Orbit Rendezvous, 10, 42
EELV (evolved expendable launch vehicle),
    37, 38
Eisele, Don, 18
electrical charge, lunar, 109
Environmental movement, 19
ESA (European Space Agency), 35, 46, 47, 55, 58, 93, 95, 115, 119,
    126, 142, 145
EVA (Extra vehicular activity), 10, 23, 25, 29, 35, 72, 74, 75, 76, 83,
    84, 96, 128, 129, 131, 145

F
Flashline Arctic Research Station, 130
Fox, Jeff, 45, 49, 50, 51, 53
Fra Mauro, 24, 31, 64, 129
Friedman, Louis, 136, 138, 139, 140
4 Frontiers Corporation, 139, 141
Fuel cells, 25, 26, 44, 66, 68

G
Galilean satellites, 106
Garvin, James, 34, 35, 41, 43, 65, 66, 99, 101, 102,
    103, 105, 106, 114, 115, 126, 128, 131,
    132, 142, 143
Gaye, Marvin, 143
Gemini, 12
Glenn Research Center, 66
Global Exploration Strategy, 115
Gloves, 52, 70, 73, 83
Golombeck, Matthew, 126
Google, 35, 59
Griffin, Mike, 37, 40, 137, 144
Grissom, Virgil "Gus", 12, 13
Grumman Aerospace, 20

H
Habitats, 55, 57, 58, 63, 68, 74, 80–92, 94, 95, 96, 110, 112, 121,
    124, 128, 129, 130, 141, 142, 145
    Inflatable, 87, 88, 89
    on ISS, 45, 85, 88
    Rigid, 86, 88
Hadley Apollo landing site, 29
Haise, Fred, 26
Hanley, Jeff, 34, 35, 40, 41, 56, 84, 133,
    138, 144
Hansen, Laurie, 54, 55, 56, 59, 60, 61
Hartmann, William K., 105
Hiten, 111
Homnick, Mark, 139, 140, 141
Houbolt, John, 10
Howe, Steven, 128
Hypergolic fuels, 57

I
IKI (Institute for Space Research, Moscow), 115–116
ILC Dover, 88–89
Ilmenite, 66
Impacts, 14, 30, 53, 66, 93, 101–103, 104, 106, 111,
    113–114, 116, 136
India, 36, 93, 116, 119
In Situ Resource Utilization (ISRU), 66–67
Intercontinental Ballistic Missiles, 5–6
International Space Station (ISS), 4–5, 43, 58, 61,
    80, 85–86, 87
Irwin, James, 29
Itokawa (asteroid), 136
Ivins, Marcia, 41, 55, 58–59, 61, 90, 96–97, 117

J
J-2/J-2X engines, 37, 38–39
Japan, 80, 111, 115–116
"J" Missions, 24, 28–30
Jodrell Bank, 8
Johns, Bill, 43, 45–46, 47, 52–53, 54, 145
Johns Hopkins University's Applied Physics
    Laboratory, 64, 115
Johnson Spaceflight Center, 27
Jules Verne Automated Transfer Vehicle, 55

K
Kaguya, 111
KIBO science laboratory, 80, 115–116
Kirasich, Mark, 44–45, 46–47, 52, 53
Komarov, Vladimir, 14
KORD system, 17, 21
Korolev, Sergei, 11, 123
Kosmo, Joe, 70, 74, 76, 117
Kretchet suit, 71
Kring, David, 80, 102, 106–107, 114–115, 116
K/T boundary, 102

L
L-1 Soviet Moon lander, 10, 11, 13–15
LaCrOSS, 112–113
Landis, Geoffrey, 126, 133, 137, 142
Langley Research Center, 91–92
Laser reflector, 108
LEDs, 72
LEO (Low Earth Orbit), 10, 35, 36, 37, 45–47,
    60, 61, 143
Leonov, Alexei, 11, 117
Levison, Hal, 100
Lewis & Clarke, 122, 144
Linkin, Vachislav, 115–116
LM (Lunar Module), 10, 19–21, 55–56, 59–60,
    83, 93, 94
Lockheed Martin Astronautics, 42–43, 46
Lovell, James, 18
Lovochkin, Semyon, 8
Low Impact Docking System (LIDS), 54
Luna Glob, 118
Luna probes, 2–4, 7–9
Lunar Orbiter, 9, 11, 16, 65, 110–111, 114, 118
Lunar Orbit Rendezvous, 10

Lunar and Planetary Lab, 80, 102–103, 112, 116
Lunar Prospector, 65–66, 108–109, 111
Lunar Reconnaissance Orbiter, 110–111, 143
Lunar rover, 24, 28, 29, 80, 104, 118, 130, 145–146
    Apollo, 24, 28, 84
    Chariot, 63, 91
    Lunakhod, 27
Lunar X-prize, 59
Luna (Soviet Moon Lander), 6–7, 8, 10–11, 21, 22, 27, 34, 59, 99–100, 105, 118
Lutz, Glen, 35, 70–71, 75, 128, 143, 145–146

M
McCandless, Bruce, 48, 60
McDivitt, James, 20
McKay, Chris, 124, 126–127, 129–131, 132, 136, 144–145
Magma seas, 100
Magnetic properties of the Moon, 108–109
Manned Maneuvering Unit (MMU), 48
Maria regions, 29, 100
Mark III spacesuit, 69, 70–71
Mars, 7, 10–11, 36, 42–43, 55, 89–90, 100–101, 102, 104, 114, 118, 120–147
Mars Direct, 125, 134
Mars Exploration Rovers, 7, 84
Marshall Space Flight Center, 37, 137, 145, 151
Mars Society, 119, 142
Mechanical counter-pressure suit, 76
Mendell, Wendell, 40, 43, 102, 103, 113, 115–116, 137
Mercury (planet), 103, 108
Mercury spacecraft, 6, 11
Meteors, 26, 63, 100, 102, 114, 126–127
Mir, 61, 117
Mishin, Vassily, 14
Mitchell, Edgar, 129
Moon dust, 107

N
N1 Soviet booster, 55
NASA (National Aeronautics and Space Administration), 4, 12, 13, 14, 18, 19, 20, 21, 34, 35, 37, 40, 41, 48, 59, 140, 144–146
National Space Society, 142
Near Earth Asteroid Rendezvous, 108
NEOs (Near Earth Objects), 140, 141
New Horizons Pluto mission, 142
North American Rockwell, 123–124
Nuclear propulsion, 128

O
Origin of the Moon, 105, 111
Orion
    Block One (near Earth), 47–48
    Block Two (lunar excursions), 47–48
    crew size, 42, 43, 46–47, 53
    flight deck, 49, 51
    reusability, 52–54
    screens, 46, 49
Orlon suit, 48

P
Paine, Thomas, 124
Parachutes, 7, 14, 42, 44, 49–50, 52, 53
Peary Crater, 63, 67
Peary, Robert, 67
Phobos Grunt mission, 127
Pioneer Astronautics, 134
Planetary Society, 58, 127, 136–137, 139, 142
PLSS lifesupport backpack, 83
Progress supply ship, 58, 122

R
Radiation, 9, 35, 63, 74, 91–92, 99, 110–111, 123
Ranger spacecraft, 6–7
Regolith, 63, 65, 66, 68, 91–92, 93, 108–109, 110–111, 115, 118, 134, 136
Renaissance, 146
Rendezvous, 10, 17, 20, 42, 47, 136
Ride, Sally, 124
RKK Energia, 58, 125–126
Robots, role with humans, 80, 133, 143
Roosa, Stuart, 28
Rosetta Stone, 99

S
Saturn V, 10, 15, 17, 18, 33–34, 37, 39, 42
Schmitt, Harrison "Jack", 83, 96, 106, 109
Schweickart, Rusty, 20
Scientific reasons to return to the Moon, 98–119
Scott, Dave, 20, 29
Shackleton Crater, 59–60, 63, 64, 67, 68, 114, 116
Shackleton, Ernest, 67
Shenzhou VI, 118
Shephard, Alan, 28, 129
Shirra, Wally, 18
SMART 1, 111
Solar arrays/solar panels, 14, 43, 44–45, 46, 63, 126, 135
Solar wind, 99, 100, 108–109, 136
Sorties, 45, 56, 75, 81, 83–84, 94, 122
South Pole-Aitken Basin, 116
Southwest Research Institute, 30, 100, 106
Soyuz, 11, 13–14, 15, 49, 55, 58, 73, 115–116
Space Exploration Initiative, 124–125
Space Science Initiative, 116
Space Shuttle, 4–5, 15–16, 36, 43, 44, 48, 68, 75, 85–86
Space Shuttle Main Engines, 37
SpaceX Corporation, 140
Spudis, Paul, 66, 112, 115, 116
Stafford, Tom, 20–21, 117
Stetson, Doug, 135–136
Stoker, Carol, 124
Streptococcus mitis, 23–24
Surveyor, 8–9, 23–24, 31, 59, 64
Swigert, Jack, 25–26

T
Telerobotics, 81–82
Temperatures, lunar, 24, 57–58
Terrestrial planets, 103

Tikonauts, 118
TMK Soviet Mars study, 123
Toups, Larry, 89–90
Transhab, 88, 125
T-tauri phase, 100
Tycho crater, 9, 30

V
Vectran, 87–88
Vehicle Assembly Building, 41
Vernadsky Institute, 118
Virgin Galactic, 35
Volcanism, 30, 100, 101, 104, 115
von Braun, Werner, 10, 121, 122, 123–124
Voskhod, 11, 14
Vostok, 11, 14

W
Water, lunar, 24, 66
Wilcox, Brian, 82, 96
Winnebago approach, 95

Y
Young, John, 20–21, 36, 72, 112

Z
Zakharov, Alexander, 127
Ziyuan, Ouyang, 118
Zond, 10–11, 17–18
Zubrin, Robert, 124, 125, 127, 134, 135,
        138, 140, 142, 144
Zvezda module, 126

Printed in the United States of America